FOCUS
REVIEW GUIDE

to accompany
Mader *Biology* AP® Edition

D1379858

Mc
Graw
Hill
Education

About the AP Consultant

Darrel James received his BS in Biology from Pacific University in Oregon and his MS in Marine Science from Oregon State. He taught Biology for thirty years, and is presently teaching at Beyer High School in Modesto, CA. Darrel teaches Pre-AP as well as AP Biology and is the Department chairperson. Darrel has served as a reader, table leader, and assistant chief reader for the AP Biology exam since 1990. He has led numerous one-day AP Biology curriculum and grading workshops for the College Board, and he has presented week-long institutes for the past twenty-four summers. Darrel is a member of the California Commission on Science and Technology. He is the vice-chair of the California Teachers Advisory Council that is an arm of the CCST which advises the state on STEM and Digitally enhanced education. Darrel has led Marine Biology trips to Hawaii for his AP Biology students. He is also an avid cyclist. He loves to teach and show people how much fun Biology can be.

COVER: Seb_c_est_bien/iStockphoto/Getty Images

mheducation.com/prek-12

Copyright © 2019 McGraw-Hill Education

Send all inquiries to:
McGraw-Hill Education
8787 Orion Place
Columbus, OH 43240

ISBN: 978-0-07-681224-0
MHID: 0-07-681224-3

Printed in the United States of America.

1 2 3 4 5 6 7 8 9 QVS 23 22 21 20 19 18

Table of Contents

Chapter	Section	AP Correlation
35 Respiratory Systems (pp. 221-225)	35.1	2.A.3, 2.D.2, 4.A.4, 4.B.2
	35.2	2.D.2, 4.A.4, 4.B.2
	35.3	2.C.1, 2.D.3, 4.B.3
36 Body Fluid Regulation and Excretory Systems (pp. 226-229)	36.1	2.D.2, 4.A.4, 4.B.2
	36.2	2.C.1, 2.D.3, 4.A.4, 4.B.2
37 Neurons and Nervous Systems (pp. 230-237)	37.1	3.E.2, 4.B.2
	37.2	3.D.2, 3.D.3, 3.E.2, 4.B.2
	37.3	3.E.2, 4.B.2
	37.4	3.D.1, 3.E.1, 3.E.2, 4.A.4, 4.B.2
38 Sense Organs (pp. 238-242)	38.1	1.C.3
	38.2	1.B.1, 1.C.3
	38.3	1.B.1, 1.C.3
	38.4	1.B.1, 1.C.3
	38.5	Extending Knowledge
39 Locomotion and Support System (pp. 243-246)	39.1	4.A.4, 4.B.2
	39.2	4.A.2, 4.A.4
	39.3	4.A.4, 4.B.2
40 Hormones and Endocrine Systems (pp. 247-251)	40.1	2.E.2, 3.D.1, 3.D.3
	40.2	2.C.1, 2.D.3, 2.E.2, 3.D.2, 3.E.2
	40.3	2.C.1, 2.C.2, 2.D.2, 2.E.2, 3.D.1, 3.D.4
41 Reproductive Systems (pp. 252-257)	41.1	2.C.1, 2.E.2
	41.2	2.C.1
	41.3	2.C.1
	41.4	2.C.1
	41.5	2.C.1
42 Animal Development and Aging (pp. 258-262)	42.1	2.E.1, 3.D.2
	42.2	2.E.1, 3.B.1, 3.B.2
	42.3	2.E.1
	42.4	Extending Knowledge

AP Unit Review (pp. 263-266)

Chapter	Section	AP Correlation
43 Behavior Ecology (pp. 267-272)	43.1	2.E.3, 3.E.1, 3.E.2
	43.2	2.C.2, 2.E.1, 2.E.2, 3.E.1
	43.3	2.C.2, 2.E.3, 3.D.1, 3.E.1
	43.4	2.C.2, 2.E.2, 2.E.3, 3.E.1
44 Population Ecology (pp. 273-280)	44.1	2.D.1
	44.2	2.D.1, 4.A.5
	44.3	2.D.1, 4.A.5, 4.B.3
	44.4	2.D.1, 2.D.3, 4.A.5, 4.B.3
	44.5	2.D.1, 4.A.5
	44.6	4.A.5, 4.A.6

Chapter	Section	AP Correlation
45 Community and Ecosystem Ecology (pp. 281-286)	45.1	2.D.1, 2.E.3, 4.A.5, 4.A.6, 4.B.3
	45.2	1.A.3, 4.B.4
	45.3	2.A.1, 2.A.3, 2.D.1, 4.A.5, 4.A.6
	45.4	4.B.3, 4.B.4
46 Major Ecosystems of the Biosphere (pp. 287-290)	46.1	4.A.6
	46.2	4.A.6
	46.3	4.A.6, 4.B.4
47 Conservation of Biodiversity (pp. 291-296)	47.1	1.A.2, 4.C.3, 4.C.4
	47.2	1.A.2, 1.C.3, 4.A.5, 4.C.3, 4.C.4
	47.3	1.C.1, 4.A.5, 4.A.6, 4.B.3, 4.B.4, 4.C.4
	47.4	4.B.4

AP Unit Review (pp. 297-300)

Using Your *Focus Review Guide*

This review guide was developed with the AP student in mind. The activities within each chapter will help you to focus on and review the key content in the chapter as it relates to the AP Biology Curriculum. Each chapter includes:

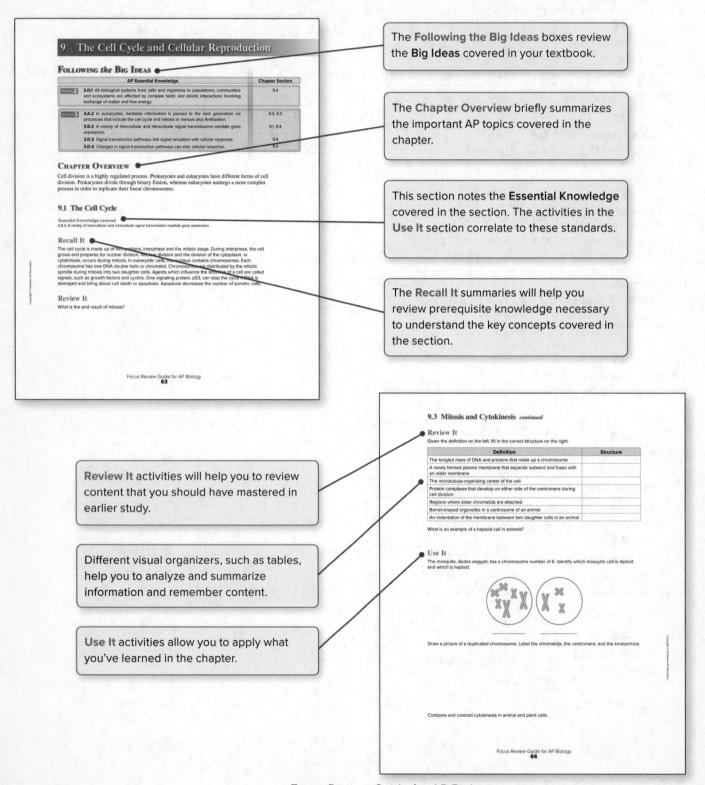

The **Following the Big Ideas** boxes review the **Big Ideas** covered in your textbook.

The **Chapter Overview** briefly summarizes the important AP topics covered in the chapter.

This section notes the **Essential Knowledge** covered in the section. The activities in the **Use It** section correlate to these standards.

The **Recall It** summaries will help you review prerequisite knowledge necessary to understand the key concepts covered in the section.

Review It activities will help you to review content that you should have mastered in earlier study.

Different visual organizers, such as tables, help you to analyze and summarize information and remember content.

Use It activities allow you to apply what you've learned in the chapter.

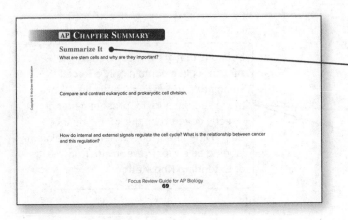

AP CHAPTER SUMMARY

Summarize It

What are stem cells and why are they important?

Compare and contrast eukaryotic and prokaryotic cell division.

How do internal and external signals regulate the cell cycle? What is the relationship between cancer and this regulation?

Focus Review Guide for AP Biology
69

Each chapter of the *Focus Review Guide* ends with a **Summarize It** section. This section contains free response questions that ask students to synthesize information from all the Big Ideas covered in the chapter.

The **AP Unit Review** provides a cumulative review of content from a set of chapters. This review with its AP-style multiple choice and free-response questions provides an excellent review of AP content while providing you an opportunity to hone your skills answering AP-style questions.

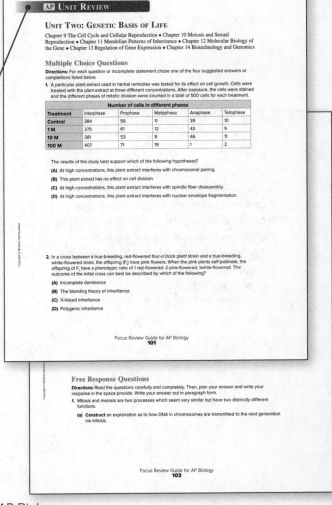

AP UNIT REVIEW

UNIT TWO: GENETIC BASIS OF LIFE

Chapter 9 The Cell Cycle and Cellular Reproduction • Chapter 10 Meiosis and Sexual Reproduction • Chapter 11 Mendelian Patterns of Inheritance • Chapter 12 Molecular Biology of the Gene • Chapter 13 Regulation of Gene Expression • Chapter 14 Biotechnology and Genomics

Multiple Choice Questions

Directions: For each question or incomplete statement chose one of the four suggested answers or completions listed below.

1. A particular plant extract used in herbal remedies was tested for its effect on cell growth. Cells were treated with the plant extract at three different concentrations. After exposure, the cells were stained and the different phases of mitotic division were counted in a total of 500 cells for each treatment.

Number of cells in different phases

Treatment	Interphase	Prophase	Metaphase	Anaphase	Telophase
Control	384	56	11	39	10
1 M	375	61	12	43	9
10 M	381	53	9	46	11
100 M	407	71	19	1	2

The results of this study best support which of the following hypotheses?

(A) At high concentrations, this plant extract interferes with chromosomal pairing.

(B) This plant extract has no effect on cell division.

(C) At high concentrations, this plant extract interferes with spindle fiber disassembly.

(D) At high concentrations, this plant extract interferes with nuclear envelope fragmentation.

2. In a cross between a true-breeding, red-flowered four-o'clock plant strain and a true-breeding, white-flowered strain, the offspring (F₁) have pink flowers. When the pink plants self-pollinate, the offspring of F₁ have a phenotypic ratio of 1 red-flowered: 2 pink-flowered; 1white-flowered. The outcome of the initial cross can best be described by which of the following?

(A) Incomplete dominance

(B) The blending theory of inheritance

(C) X-linked Inheritance

(D) Polygenic inheritance

Focus Review Guide for AP Biology
101

Free Response Questions

Directions: Read the questions carefully and completely. Then, plan your answer and write your response in the space provide. Write your answer out in paragraph form.

1. Mitosis and meiosis are two processes which seem very similar but have two distinctly different functions.

(a) **Construct** an explanation as to how DNA in chromosomes are transmitted to the next generation via mitosis.

Focus Review Guide for AP Biology
103

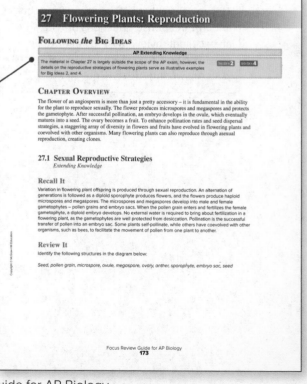

30.3 Evolution of Early Genus *Homo* continued

Use It

Describe the origins and speciation of *Homo floresiensis*.

30.4 Evolution of Later Genus *Homo*
Extending Knowledge

Recall It

The replacement model or out-of-Africa-hypothesis is the most widely accepted hypothesis for the evolution of modern humans. This hypothesis proposes the modern human evolved from earlier *Homo* species found only from Africa, before migrating to Asia and Europe and displacing the other *Homo* species found there. This would have included the species *Homo neandertalensis* whose fossils have been found throughout Europe. Although, interbreeding may have occurred between *Homo neandertalensis* and *Homo sapiens* (modern humans). Cro-Magnons are the oldest *Homo sapiens*. Modern *Homo sapiens* are quite diverse externally as different populations evolved as adaption to local conditions but genetically are very similar.

Review It

Fill in the most probable word or description associated with the later genus *Homo*.

Term	Description
biocultural evolution	
	Oldest fossils of *Homo sapiens*
Denisovans	
hunter-gatherers	
	archaic humans that lived 200,000-300,000 years ago

Modern humans have a wide variation in phenotype. What are two hypotheses that explain these variations?

While populations of humans across the world do show variation in phenotypes, what evidence is there that we all arose from a common ancestor?

Sections that are not covered by the AP Curriculum contain only a **Recall It** summary and **Review It** activities. While these sections do not present material directly covered on the AP Exam, they do provide important content. These are labeled as either **Prerequisite Knowledge** or **Extending Knowledge**.

Some chapters are beyond the scope of the AP curriculum and the AP Exam. These chapters start with the **AP Extending Knowledge** table, and contain questions that ask students to apply the Big Ideas to new situations.

27 Flowering Plants: Reproduction

FOLLOWING *the* BIG IDEAS

AP Extending Knowledge
The material in Chapter 27 is largely outside the scope of the AP exam, however, the details on the reproductive strategies of flowering plants serve as illustrative examples for Big Ideas 2, and 4.

CHAPTER OVERVIEW

The flower of an angiosperm is more than just a pretty accessory – it is fundamental in the ability for the plant to reproduce sexually. The flower produces microspores and megaspores and protects the gametophyte. After successful pollination, an embryo develops in the ovule, which eventually matures into a seed. The ovary becomes a fruit. To enhance pollination rates and seed dispersal strategies, a staggering array of diversity in flowers and fruits have evolved in flowering plants and coevolved with other organisms. Many flowering plants can also reproduce through asexual reproduction, creating clones.

27.1 Sexual Reproductive Strategies
Extending Knowledge

Recall It

Variation in flowering plant offspring is produced through sexual reproduction. An alternation of generations is followed as a diploid sporophyte produces flowers, and the flowers produce haploid microspores and megaspores. The microspores and megaspores develop into male and female gametophytes – pollen grains and embryo sacs. When the pollen grain enters and fertilizes the female gametophyte, a diploid embryo develops. No external water is required to bring about fertilization in a flowering plant, as the gametophytes are well protected from desiccation. Pollination is the successful transfer of pollen into an embryo sac. Some plants self-pollinate, while others have coevolved with other organisms, such as bees, to facilitate the movement of pollen from one plant to another.

Review It

Identify the following structures in the diagram below:

Seed, pollen grain, microspore, ovule, megaspore, ovary, anther, sporophyte, embryo sac, seed

1 A View of Life

FOLLOWING *the* BIG IDEAS

AP Essential Knowledge	Chapter Section
BIG IDEA 1 — **Big Idea 1** The process of evolution drives the diversity and unity of life.	1.2
BIG IDEA 2 — **Big Idea 2** Biological systems utilize free energy and molecular building blocks to grow, to reproduce and to maintain dynamic homeostasis.	1.3
BIG IDEA 3 — **Big Idea 3** Living systems store, retrieve, transmit and respond to information essential to life processes.	1.4
BIG IDEA 4 — **Big Idea 4** Biological systems interact, and these systems and their interactions possess complex properties.	1.5

CHAPTER OVERVIEW

This chapter introduces you to the four Big Ideas of the AP Biology curriculum: Big Idea 1, Evolution; Big Idea 2, Energy and Molecular Building Blocks; Big Idea 3, Information Storage, Transmission, and Response; and Big Idea 4, Interdependent Relationships. This chapter provides an overview of the Big Ideas, and how they are interconnected to provide an in-depth and fundamental study of biology. This chapter also outlines the seven science practices that you will master as you learn to think like a scientist.

1.1 Introduction to AP Biology
Extending Knowledge

Recall It

Biology is the study of life on Earth. There are several basic characteristics shared by all living organisms. Living organisms possess organized systems, and require energy and materials to maintain a stable internal environment. All living things reproduce, develop and grow, and respond to stimuli. Living organisms adapt to changing conditions.

Review It

In your own words, what is the field of biology all about?

1.1 Introduction to AP Biology *continued*
Extending Knowledge

In the chart below, describe the attributes of living things.

Place a star next to the category that all living things have in common:

Size	
Locations	
Life span	
Composition	

1.2 Big Idea 1: Evolution

Recall It

Evolution is the core concept of biology. Evolution is the graduate change in populations of organisms over time. Natural selection, described by Charles Darwin, is the mechanism by which evolution progresses. Organisms are classified through taxonomy, and the evolutionary relationships are described by systematics.

Review It

List the basic classification of organisms going from most inclusive to least inclusive:

Use It

Explain the phrase "common descent with modification."

1.3 Big Idea 2: Energy and Molecular Building Blocks

Recall It

All biological processes are driven by the input of energy from many sources. Organisms require energy to grow, reproduce, and maintain organization, and have developed complex mechanisms for energy capture and storage. Living systems maintain their ideal temperature, moisture level, acidity, and other internal physiological factors through homeostasis.

Review It

Why is the sun the ultimate source of energy for nearly all life on Earth?

1.3 Big Idea 2: Energy and Molecular Building Blocks *continued*

Use It

Draw a simplistic diagram to describe how chemicals and energy flow from the sun to plants to organisms that eat the plants.

1.4 Big Idea 3: Information Storage, Transmission, and Response

Recall It

From chemicals and DNA at the molecular level to complex behaviors, such as predation or mating dances, information runs the world of biology. All life must reproduce, grow and develop, adapt to change, and respond to stimuli. Storing, transmitting, and responding to information is critical for all life processes. DNA provides the instructions for the organization for each particular organism, and mutations in genetic variation provide the possibility for change in its species. Sometimes, organisms inherit characteristics that afford them a greater chance of survival.

Review It

At the molecular level, where is the fundamental information for all of life's processes stored?

Use It

How do you communicate an idea you've had to a friend, and how does your friend receive that message?

1.5 Big Idea 4: Interdependent Relationships

Recall It

In living systems, cooperation takes place at multiple levels. From molecules to cells to populations in an ecosystem, living things interact and create emergent properties. Damaging or destroying any part of a living system will affect the whole as the parts are all tightly interconnected. Any place there can be more diversity or more options in a living system, the better change life has of survival as complexity allows for flexibility to change.

1.5 Big Idea 4: Interdependent Relationships *continued*

Review It

All systems are made up of intricate parts. Place the following emergent properties in order from least complex to most complex: *organs, ecosystems, cells, tissues, organisms*

Use It

In the chart below, briefly describe how the two terms interact:

Organisms		Ecosystems
Cells		Organs
Autotrophs		Heterotrophs

1.6 The AP Science Practices and the Process of Science

Recall It

The scientific method outlines the way that scientists conduct experiments and observational studies. In AP Biology, you will use the seven science practices as tools during your investigations. Like the scientific method, the science practices are applicable across all fields of science.

Review It

Describe the seven science practices (SP) that will help you understand biology and how scientists approach their work:

SP 1	
SP 2	
SP 3	
SP 4	
SP 5	
SP 6	
SP 7	

1.6 The AP Science Practices and the Process of Science *continued*

In the flow chart below, **identify** the process of the scientific method missing needed to conclude which hypothesis is correct:

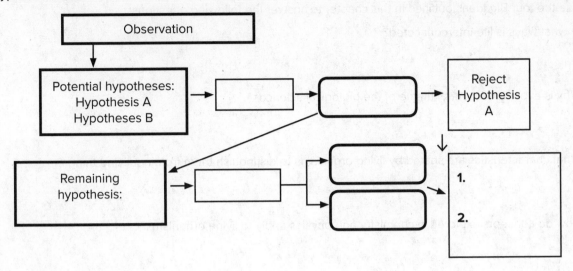

Use It

Find **Figure 1.13** in your textbook. The graph at the bottom shows how effective a particular treatment was against a disease in three different groups. Using this graph, answer some questions concerning the scientific method:

What question might the scientists have asked before starting this study?

What might have been a hypothesis for this study?

Why did the scientists include a control group in their experimental design?

Explain what the "T" shaped lines coming off of the bars on the graph indicate.

If the scientist who performed a statistical test on this data found a p value less than 0.05 between Group 1 and Group 2, is there a significant difference between the two groups?

These scientists sent their study and conclusions off to a scientific journal. What happens to their article before it is published?

Summarize It

Use the four Big Ideas outlined in this chapter to answer the following questions:

In what ways is life interconnected?

Why is evolution a central theme of the biological sciences?

What characteristics are shared by living organisms to distinguish them from nonliving things?

How do concepts explored in chemistry and physics apply to living organisms?

What are the science practices, and how are they different than the four Big Ideas of AP Biology?

2 Basic Chemistry

FOLLOWING *the* BIG IDEAS

AP Essential Knowledge	Chapter Section
BIG IDEA 2 2.A.3 Organisms must exchange matter with the environment to grow, reproduce, and maintain organization	2.3
BIG IDEA 4 4.A.1 The subcomponents of biological molecules and their sequence determine the properties of that molecule.	2.2

CHAPTER OVERVIEW

To fully understand biology, it is important to also study chemistry and physics as matter and free energy are required for organisms to survive. All living things are composed of matter. Carbon, hydrogen, oxygen, nitrogen make up the majority of matter in living things. These elements, along with phosphorus, and sulfur, and a few other trace elements are required for life. Elements are substances that can't be broken down into simpler substances (by ordinary chemical means). Elements are composed of atoms, which in turn are comprised of protons, neutrons, and electrons. This chapter reviews how atoms transfer and share electrons, how molecules are formed, and the different types of bonds that exist. This chapter also describes the importance of the unique chemistry of water.

2.1 Chemical Elements
Extending Knowledge

Recall It

Anything that takes up space and has mass is called matter. Everything nonliving and living on this earth is composed of basic substances called elements. These basic substances are made up of tiny particles we call atoms, and they, in turn, contain subatomic particles including protons, electrons and neutrons. The number of protons housed in the nucleus is called the atomic number. The sum of the sum of the number of protons and neutrons in the nucleus is called the mass number. Atoms of the same element that have different numbers of neutrons are called isotopes.

Review It

On the periodic table, elements are listed with two numbers and one symbol. Define what each means.

   Atomic number:

   Atomic symbol:

   Atomic mass:

Some isotopes are highly radioactive. What is radiation? Describe one way radiation is harmful and one way it is beneficial to humans.

2.1 Chemical Elements *continued*
Extending Knowledge

A common model used to illustrate electrons in electron shells around the nucelus are called Bohr models. This model is described on page 23. Following the rules of the Bohr model, draw a model of boron. Label the *nucleus*, *electrons*, and *valance shell*.

2.2 Molecules and Compounds

Essential Knowledge covered
4.A.1: The subcomponents of biological molecules and their sequence determine the properties of that molecule.

Recall It

When atoms bond together, a molecule is formed. The combination of two or more different elements is referred to as a compound. Molecules and compounds are described by formulas, which indicate number and types of atoms present. Electrons play a critical role in the bond formation between elements. An element that has lost or gained an electron from its valence shell is called an ion. Ionic bonds are held together by an attraction between negatively and positively charged ions; whereas covalent bonds share electrons to satisfy their valence shell's octet. If you were to measure an atom's attraction for an electron, you would be measuring its electronegativity. Covalent bonds that share electrons equally are referred to as nonpolar covalent, and covalent bonds which share unequally are called polar covalent. Determined by both bond polarity and by molecular shape, the polarity of a molecule affects molecule-molecule interactions.

Review It

Describe the difference between ionic, polar covalent, and nonpolar covalent bonds. Use the following keywords in your description: ions, electrons, and electronegativity.

Draw an arrow to the bonds which hold these molecules together. Classify the bonds of the molecules as either ionic (I) or covalent (C) in the table below.

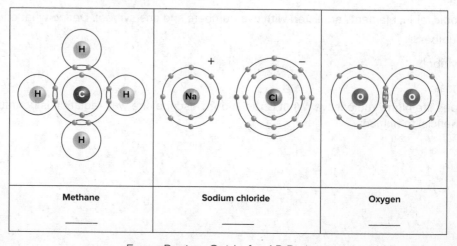

Methane	Sodium chloride	Oxygen
_____	_____	_____

2.2 Molecules and Compounds *continued*

Use It

Students studying for their AP Exams might find themselves imbibing $C_8H_{10}N_4O_2$, also known as caffeine. **Identify** the number of atoms and molecules in this compound.

Caffeine, $C_8H_{10}N_4O_2$, is made up of many different covalent bonds between its atoms. Given the information that the nitrogen and oxygen atoms present have higher electronegativity compared to the carbon and hydrogen atoms, **determine** whether or not caffeine is polar and describe why.

2.3 Chemistry of Water

Essential Knowledge covered
2.A.3: Organisms must exchange matter with the environment to grow, reproduce, and maintain organization

Recall It

Water is an amazing molecule. In order for life on earth to exist and continue existing, life depends on the structure and properties of water, along with its behavior. Water is formed from oxygen and hydrogen atoms through highly polar covalent bonds. The partial charges formed at both the oxygen and hydrogen ends of the molecule allow for chemical associations known as hydrogen bonding. The nature of these bonds are weaker than covalent bonds, but the sheer abundance of these bonds allow for water's many unique physical properties. Water molecules are cohesive, or attracted to other water molecules. Cohesion helps to facilitate the movement and exchange of water between an organism and its environment. Water is also adhesive, creating hydrogen bonds with other polar substance. Adhesion between water and other polar substances allows for phenomena such as capillary action.

Review It

What is hydrogen bonding?

Draw a simple diagram of how water travels through a tree. Label where adhesion and cohesion occurs.

2.3 Chemistry of Water *continued*

Use It

List and describe the properties of water which allows each organism in the chart below their certain ability.

Organism	Ability	Water Property	How it works
Water strider	Walks across the surface of a pond		
Oak tree	Photosynthesizes 25 m above ground		
Human	Cools off on a hot day by jumping in the ocean		
Lake trout	Over-winters in lake that is covered in 10 cm of solid ice		

The graph below shows the energy required to change the temperature of a gram of water. Using the concept illustrated on graph below, **explain** how a camel can live in a desert where temperatures changing from 5 to 40 °C over the course of 24 hours.

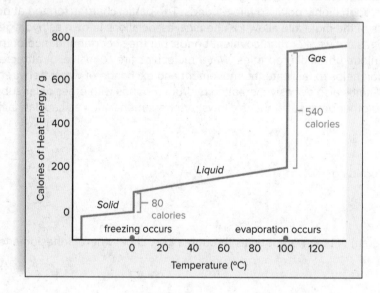

Describe how water can be used as a solvent to wash road salt, $CaCl_2$, off of a car.

Recall It

A solution's pH is determined by its concentration of hydrogen ions. A solution with a pH lover than 7 is called an acid. A solution with a pH higher than 7 is called a based. And a solution with a pH of 7 is neutral. A buffer is a chemical that helps keep solutions within a specific range of a pH. Buffers release hydrogen ions when a base is added to a solution and absorb hydrogen ions when an acid is added. Buffers are very important in maintaining constant pH's of biological systems. Blood is buffered by carbonic acid by taking up excess OH^- or H^+ ions.

Review It

Describe what makes a chemical an acid or a base.

Identify if these solutions have more hydrogen ions, more hydroxide ions, or an equal number of hydrogen and hydroxide ions based on their pH.

Solution	pH	Ion
Coffee	5	
Tears	7.0	
Soda	3.0	
Bleach	12.0	

AP CHAPTER SUMMARY

Summarize It

In the following molecule, which bond is the strongest and why?

$$O=OH-C\equiv C-HH-H$$

Why is hydrogen bonding one of the most important bonds, even though it's considered a 'weak' bond? And where can we find hydrogen bonds in nature?

3 The Chemistry of Organic Molecules

FOLLOWING *the* BIG IDEAS

	AP Essential Knowledge	Chapter Section
BIG IDEA 2	**2.A.3** Organisms must exchange matter with the environment to grow, reproduce, and maintain organization.	3.1
	2.B.1 Cell membranes are selectively permeable due to their structure.	3.3
BIG IDEA 3	**3.A.1** DNA, and in some cases RNA, is the primary source of heritable information.	3.5
BIG IDEA 4	**4.A.1** The subcomponents of biological molecules and their sequence determine the properties of that molecule.	3.1, 3.2, 3.3, 3.4, 3.5

CHAPTER OVERVIEW

The macromolecules of life include: carbohydrates, lipids, proteins, and nucleic acids. These macromolecules are combined to produce larger structures, which lead them to have different functions. Energy storage, data storage, and all biological processes are the result of these functions. Biological molecules maintain the essential structures of living organism, including cell membranes.

3.1 Organic Molecules

Essential Knowledge covered
4.A.1: The subcomponents of biological molecules and their sequence determine the properties of that molecule.

Recall It

Carbon is the essential building block of life. How and what atoms are arranged around carbon leads to astonishing diversity in organic compounds. Functional groups are specific combinations of bonded atoms that account for differences in chemical properties of organic molecules. Some molecules, known as isomers, have the same molecular formula but different structures. Changes in the structure of a molecule can have great effects on biological function. Biological macromolecules such as carbohydrates, nucleic acids, proteins, and lipids allow organisms to grow, reproduce, and maintain organization. The long chains of similar subunits in macromolecules are joined by dehydration reactions (condensation) and are broken down by hydrolysis reactions.

Review It

Complete the following chart of biomolecule monomer and polymer makeup:

Biomolecule	Monomers	Polymer
	Amino Acids	
Carbohydrate		Polysaccharide
		DNA, RNA
Lipids	Glycerol, fatty acids	

3.1 Organic Molecules *continued*

Carbon has unique properties which allow for many different types of molecules to be formed from it. Describe how each property allows for the formation of different biomolecules.

Property	
Number of Electrons	
Shape	
Bonds	

Use It

Determine whether or not the following statements are true or false (T/F) about functional groups:

R describes where the functional group attaches to a carbon skeleton.

A functional group determines an organic molecule's polarity.

A functional group does **not** determine the types of reactions an organic molecule will undergo.

The carbon skeleton acts as a framework to position the functional group.

3.2 Carbohydrates

Essential Knowledge covered
4.A.1: The subcomponents of biological molecules and their sequence determine the properties of that molecule.

Recall It

Carbohydrates are made up of carbon and hydrogen, and include both single sugar molecules and chains of sugars. Single, or monomer, subunits are called monosaccharaides, and long chains, or polymers, are called polysaccharides. A monosaccharide with six carbons is called a hexose. The hexose glucose is the most common form of monosaccharide cellular fuel. Glucose is stored as starch in plants and as glycogen in animals. With five carbons, the pentose sugars ribose and deoxyribose are present in RNA and DNA. Two monosaccharaides joined through dehydration is called a disaccharide.

Review It

What are the two major roles that carbohydrates play in living organisms?

Name three structural polysaccharides described in this section.

3.2 Carbohydrates *continued*

Use It

These carbohydrates are illustrated in your textbook on the following page numbers. Examine the structure of each carbohydrate and classify the carbohydrate as being a monosaccharide (M), disaccharide (D), or a polysaccharide (P).

Carbohydrate	Illustration Page #	Structure
Glucose	40	
Starch	41	
Maltose	41	
Cellulose	42	
Glycogen	41	

3.3 Lipids

Essential Knowledge covered
4.A.1: The subcomponents of biological molecules and their sequence determine the properties of that molecule.

Recall It

Lipids are important to energy storage and structure in living organisms. Lipids are insoluble in water due to their nonpolar hydrocarbon chains. Triglycerides are lipids made up of three fatty acids linked to a single three-carbon glycerol, and store energy very efficiently. Phospholipids contain fatty acids, glycerol, and a polar phosphate group. Phospholipids form membranes. When placed in water, phospholipids form lipid bilayers, two layers of hydrophilic phosphate heads pointing out, and hydrophobic tails pointing in. There are also steroids and waxes. Steroids are lipids, such as cholesterol, that provide structural support in membranes, and serve as a precursor to other hormones. Wax helps prevent desiccation and mitigate bacterial growth.

Review It

Identify the lipid that plays each function in the organisms listed below:

Organism	Function	Lipid
Succulent plant	Helps keep water in leaves during hot desert days	
Sunflower	Helps store energy in seeds	
Penguin	Helps bird withstand the cold Antarctic	
Peacock	Determines difference between males and females	
All organisms	Maintains structure and function of a cell	

Compare the functions of triglycerides and phospholipids in biological systems.

Triglycerides	Phospholipids

3.3 Lipids *continued*

Use It

Draw a diagram of 12 phospholipids in a lipid bilayer in water. Label the hydrophilic and hydrophobic ends.

Why are biological membranes not formed out of triglycerides?

3.4 Proteins

Essential Knowledge covered
4.A.1: The subcomponents of biological molecules and their sequence determine the properties of that molecule.

Recall It

Proteins are behind most structure and function in cells. Proteins are built from amino acid monomers. Amino acids are joined through the processes of condensation which forms a peptide bond. If more than two amino acids are bonded together, the resulting molecule is a polypeptide. Proteins fold into special structures and become denatured when they lose their shape. Misfolded proteins that cause disease are called prions.

Review It

Describe three functions that proteins play in the cell of a living organism.

Identify the level of protein organization given its definition:

Definition	Structure
The folding that results in the final-three dimensional shape of a polypeptide	
The linear sequence of amino acids encoded for by DNA	
The coiling or folding (such as α helices or β sheets) of a polypeptide	
The interaction of two or more folded polypeptides.	

3.4 Proteins *continued*

Use It

Draw a dipeptide composed of two generalized amino acids. Label the amino end, the carboxyl end, and the R groups.

What type of reaction has to have occurred in order for these two amino acids to bond?

Chaperone proteins help proteins fold into their correct shape. If chaperone proteins are missing or nonfunctional in a cell, how might this affect a human?

3.5 Nucleic Acids

Essential Knowledge covered
3.A.1: DNA, and in some cases RNA, is the primary source of heritable information.
4.A.1: The subcomponents of biological molecules and their sequence determine the properties of that molecule.

Recall It

DNA and RNA are nucleic acids, which encode all the information needed for life. DNA stores genetic information, while RNA carries the information from DNA, facilitates protein synthesis, and may even be involved in gene regulation. Nucleic acids are composed of nucleotides. Each nucleotide has a five carbon sugar (ribose in RNA or deoxyribose in DNA), a phosphate group, and a nitrogen base. There are five types of nitrogen bases: adenine, thymine, guanine, cytosine, and uracil. Uracil is only found in the nucleic acid RNA in place of thymine. Nucleotides can be classified as purines or pyridines based on their single or double ring structure.

Review It

Name two functions that DNA has in a cell.

Identify these three important structures given the following information.

Structure	Nitrogen base(s)	Pentose sugar
	adenine, guanine, thymine, cytosine	deoxyribose
	adenine, guanine, uracil, cytosine	ribose
	adenine	ribose

3.5 Nucleic Acids *continued*

Identify the three molecules that make up a nucleotide, using the following illustration for guidance:

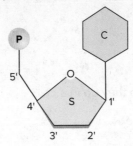

Use It

Write the missing complementary base pairs for the following DNA sequence:

5'	T	T	A	___	___	3'
3'	___	___	___	G	C	5'

If this strand was RNA, how would it be different?

Identify which bases are pyrimidines and which are purines, and how they differ in structure.

What type of bond holds complementary base pairs together?

AP CHAPTER SUMMARY

Summarize It

Using Table 3.1 in your textbook, describe two different functional groups. Where are they found and why are they significant? What do all functional groups have in common?

Cellulose and starch are both made out of glucose and both are found in plants. However, humans can metabolize starch but not cellulose. What is the difference between these two carbohydrates that causes this?

What are three types of RNA and what do they do in a cell?

Which function is more probable for the following nucleotide sequence?

GCTCCAGGTCA
(a) to store heritable information

or

(b) to carry genetic information
Justify your answer.

4 Cell Structure and Function

FOLLOWING *the* BIG IDEAS

	AP Essential Knowledge	Chapter Section
BIG IDEA **1**	**1.B.1** Organisms share many conserved core processes and features that evolved and are widely distributed among organisms today.	4.3, 4.5, 4.7, 4.8
BIG IDEA **2**	**2.A.3** Organisms must exchange matter with the environment to grow, reproduce, and maintain organization.	4.1
	2.B.3 Eukaryotic cells maintain internal membranes that partition the cell into specialized regions.	4.2, 4.3, 4.4, 4.5, 4.6, 4.7, 4.8
BIG IDEA **3**	**3.A.3** The chromosomal basis of inheritance provides an understanding of the pattern of passage (transmission) of genes from parent to offspring.	4.5
BIG IDEA **4**	**4.A.2** The structure and function of subcellular components, and their interactions, provide essential cellular processes.	4.1, 4.2, 4.3, 4.4, 4.4, 4.5, 4.6, 4.7, 4.8
	4.B.2 Cooperative interactions within organisms promote efficiency in the use of energy and matter.	4.3, 4.4, 4.5, 4.6, 4.7, 4.8

CHAPTER OVERVIEW

Cells are the fundamental building blocks of life. There are two types of cells: prokaryotes and eukaryotes. These two cell types encompass three domains: archaea, prokaryotes, and eukaryotes. While different cell types have distinguishing characteristics, all cells contain DNA and are able to carry out the metabolic processes necessary for life.

4.1 Cellular Level of Organization

Essential Knowledge covered
2.A.3: Organisms must exchange matter with the environment to grow, reproduce, and maintain organization.
4.A.2: The structure and function of subcellular components, and their interactions, provide essential cellular processes.

Recall It

The smallest unit of living matter is a cell. Cell theory, developed from the work of scientists Schleiden, Schwann, and Virchow, states: 1) All organisms are composed of cells, 2) Cells are the basic units of structure and function in organisms, and 3) Cells come only from preexisting cells because cells are self-reproducing. The surface-area-to-volume ratio limits the size of a cell because it needs a surface area large enough to allow for nutrient absorption and for waste elimination.

Review It

Name two scientists who helped determine that plants and animals are made up of cells or who played a role in developing the cell theory.

4.1 Cellular Level of Organization *continued*

Review Figure 4.2 on page 59 to describe the size of various objects and what instruments we need to observe each (E = eye, LM = light microscope, EM = electron microscope).

Living Organism or Component	Approximate Size (metric unit)	Instrument Needed to Observe
Human		
Human Egg		
Human Cell		
Bacterial Cell		
Virus		
Protein		
Amino Acid		
Atom		

Use It

Which figure has more surface area per volume?

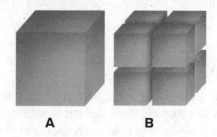

A   B

If these cubes were cells, which would be better at exchanging molecules, such as nutrients or waste with the outside environment and why?

4.2 Prokaryotic Cells

Essential Knowledge covered
2.B.3: Eukaryotic cells maintain internal membranes that partition the cell into specialized regions.
4.A.2: The structure and function of subcellular components, and their interactions, provide essential cellular processes.

Recall It

There are two cell types, prokaryotes and eukaryotes. Prokaryotic cells compose the domains Eubacteria and Archaea. All cell types share a plasma membrane and contain DNA and RNA but they differ greatly in their size and organization. Prokaryotic cells are relatively simple compared to eukaryotic cells. Prokaryotic cells lack a nucleus, an internal membrane system, and membrane-bound organelles. Prokaryotes have three basic shapes: bacillus, coccus, and sprilla or spirochetes. In most prokaryotes, DNA is organized into a single circular chromosome. All prokaryotic cells have ribosomes and carry out processes that eukaryotic cells do, such as transcription and translation. In most bacterial prokaryotes, the rigid cell wall surrounding the plasma membrane is composed of peptidoglycan. Outside the cell, prokaryotes can have external structures, such as flagella that allow them to move, fimbiae that allow them to stick to things, or conjugation pili that allow for the movement of DNA from cell to cell.

4.2 Prokaryotic Cells *continued*

Review It

In this section, you were introduced to three different types of cells: prokaryotic, eukaryotic, and archaens. What makes them similar? What makes them different?

	Shared Traits	Unique Traits
Prokaryote		
Eukaryote		
Archaens		

List three structures found in all prokaryotes.

Use It

Draw a diagram of a bacillus prokaryote. Include and identify the following structures on your diagram:

Structures:
- Flagellum
- Nucleoid
- Plasma Membrane
- Ribosome
- Cell Wall
- Fimbriae
- Mesosome

Compare and contrast prokaryotic and eukaryotic ribosomes.

Describe the role of thylakoids in cyanobacteria.

4.3 Introduction to Eukaryotic Cells

Essential Knowledge covered
1.B.1: Organisms share many conserved core processes and features that evolved and are widely distributed among organisms today.
2.B.3: Eukaryotic cells maintain internal membranes that partition the cell into specialized regions.
4.A.2: The structure and function of subcellular components, and their interactions, provide essential cellular processes.
4.B.2: Cooperative interactions within organisms promote efficiency in the use of energy and matter.

Recall It

Eukaryotic cells are more complex than prokaryotic cells and contain organelles. According to the endosymbiotic theory, it is thought that eukaryotic cells arose from the engulfment of a smaller prokaryotic cell by a larger cell. Eukaryotic cells remain relatively small, which allows materials to be moved across the membrane with ease. Eukaryotic cells have a membrane-bound nucleus, an endomembrane system, and many different membrane-bound organelles that carry out specialized functions. The nucleus is the information center of the cell. It is surrounded by two phospholipid bilayers, the outer that is continuous with the endomembrane system. Ribosomes are assembled in the nucleolus and then sent to the cytoplasm, where they translate mRNA to produce polypeptides. Vesicles are membranous sacs that transport molecules around inside the cytoplasm.

Review It

Identify the three major components of a eukaryotic cell:

Definition	Structure
An extensive network of protein fibers which maintains cell shape and assists with cell movement	
The compartments of a eukaryotic cell that carry out specialized functions	
Membranes sacs that enclose molecules and are transported around the cell	

Name one of the major organelles that is found in plants but not in animal cells. What function does this organelle give plants that animals lack?

Use It

Describe how the cytoskeleton is used to send a message from the endoplasmic reticulum to the Golgi apparatus.

Describe two reasons why internal membranes are so important in eukaryotic cells.

4.4 The Nucleus and Ribosomes

Essential Knowledge covered
2.B.3: Eukaryotic cells maintain internal membranes that partition the cell into specialized regions.
3.A.3: The chromosomal basis of inheritance provides an understanding of the pattern of passage (transmission) of genes from parent to offspring.
4.A.2: The structure and function of subcellular components, and their interactions, provide essential cellular processes.

Recall It

The nucleus of the cell contains the genetic information passed on from generation to generation, the instructions for copying itself (replication), and the information (transcription) that ribosomes use to carry out protein synthesis (translation). Genes are the carriers of genetic information, organized on chromosomes. Ribosomes are assembled in the nucleolus of the nucleus and then sent through nuclear pores of the nuclear envelope to the cytoplasm, where they translate mRNA to produce polypeptides. Three types of RNA are produced in the nucleolus: mRNA, rRNA, and tRNA. Signal peptides cause ribosomes to bind to the endoplasmic reticulum where the protein is folded into its final shape.

Review It

Provide a definition for the following parts of the nucleus:

Structure	Definition
Nucleolus	
Nuclear envelope	
Nuclear pores	
Chromosomes	
Chromatin	

Place the following events in order (1–4).

mRNA is copied from a gene and exits the nucleus

The SRP brings the mRNA to the rough ER and a polypeptide is synthesized

A ribosome attaches to the mRNA and begins protein synthesis, producing a signal protein (SRP)

An enzyme removes the SRP and the protein is folded

Use It

Why is the nucleus important to cell structure and function?

How does a message intended for another part of the cell get in and out of the nuclear envelope?

4.4 The Nucleus and Ribosomes *continued*

Identify the following structures on the nucleus illustrated:

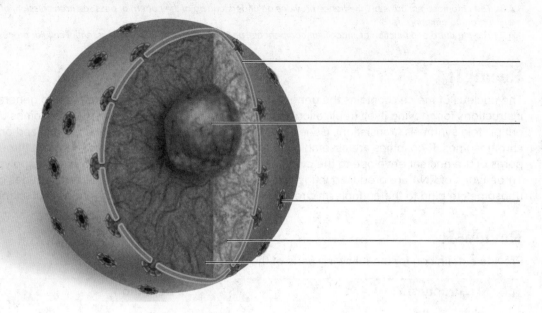

4.5 The Endomembrane Structure

Essential Knowledge covered
1.B.1: Organisms share many conserved core processes and features that evolved and are widely distributed among organisms today.
2.B.3: Eukaryotic cells maintain internal membranes that partition the cell into specialized regions.
4.A.2: The structure and function of subcellular components, and their interactions, provide essential cellular processes.
4.B.2: Cooperative interactions within organisms promote efficiency in the use of energy and matter.

Recall It

The endomembrane system divides the cell into compartments. The endomembrane system is made up of the following parts: endoplasmic reticulum (ER), the Golgi apparatus, and vesicles, including lysosomes. All compartments work together to ensure the cells function efficiently. The ER creates channels and passages within the cytoplasm. There are two types of ER, the rough ER (RER), and the smooth endoplasmic reticulum (SER). The RER is a site of protein synthesis (transcription); it also modifies proteins, and manufactures membranes. It is "rough" because it is studded with ribosomes. The SER has many roles, including involvement in carbohydrate and lipid synthesis and in detoxification. The function of the Golgi apparatus is to sort and package proteins that it receives from the ER via vesicles. It then modifies, packages, and repackages these macromolecules for export to other parts of the cell. Lysosomes contain digestive enzymes, which break down macromolecules and recycle the components of old organelles.

4.5 The Endomembrane Structure *continued*

Review It

Provide the name or the function of the following organelles or structures:

The Endomembrane System	
Structure	**Function**
	A complicated system of membranous channels and saccules
The Golgi apparatus	
	Membrane-bound vesicles produced by the Golgi apparatus that aid in digesting cellular material
	A double membrane around the nucleus that separates it from the cytoplasm

Compare and contrast the smooth endoplasmic reticulum with the rough endoplasmic reticulum.

Smooth ER	Rough ER

Both ERs

Use It

A rare inherited disorder called Tay-Sachs disease is characterized by a progressive degeneration of the nervous system due to a missing lysosomal enzyme. How are nerve cells affected by the loss of this enzyme?

Illustrate and describe how a transport vesicle enters the cis face of the Golgi apparatus and exits through the trans face to secrete its contents outside of the cell.

4.6 Microbodies and Vacuoles

Essential Knowledge covered
2.B.3: Eukaryotic cells maintain internal membranes that partition the cell into specialized regions.
4.A.2: The structure and function of subcellular components, and their interactions, provide essential cellular processes.
4.B.2: Cooperative interactions within organisms promote efficiency in the use of energy and matter.

Recall It

Microbodies are membrane-bound vesicles that contain specialized enzymes that perform specific metabolic functions. Peroxisomes enclose enzymes that break down fatty acids. Vacuoles are larger than vesicles and aid in the storage of substances. Vacuoles are essential to plant function. Plant cells typically have a large central vacuole that is filled with a watery fluid. The central vacuole maintains turgor pressure in the plant, which provides structural support.

Review It

Fill in the blanks in the sentences below with the correct organelle:

Membranous sacs called _____ are used for storing and breaking down waste.

All _____ contain enzymes whose reaction result in the production of hydrogen peroxide.

How does the central vacuole provide structural support to the cell?

Use It

In what way is the central vacuole in plants analogous to the lysosomes in animal cells?

Compare and contrast a peroxisome and a vacuole.

4.7 The Energy-Related Organelles

Essential Knowledge covered
1.B.1: Organisms share many conserved core processes and features that evolved and are widely distributed among organisms today.
2.B.3: Eukaryotic cells maintain internal membranes that partition the cell into specialized regions.
4.A.2: The structure and function of subcellular components, and their interactions, provide essential cellular processes.
4.B.2: Cooperative interactions within organisms promote efficiency in the use of energy and matter.

Recall It

The mitochondria and chloroplasts are organelles that are involved in energy production. They are similar in that they have a double-membrane structure, contain their own DNA, and can divide independently. They are different in overall structure and process. Mitochondria have an extensively folded inner membrane called cristae. On the surface and in the cristae, proteins carry out cellular respiration metabolism of sugar to produce ATP (sugar + $O_2 \rightarrow CO_2 + H_2O$ + energy). Chloroplasts, on the other hand, generate ATP by capturing light energy via thylakoid membranes through photosynthesis (solar energy + $CO_2 + H_2O \rightarrow$ sugar + O_2).

The thylakoid membranes are arranged in stacks called grana. Chloroplasts are a type of plastid, a specialized plant organelle that can have varied functions.

4.7 The Energy-Related Organelles *continued*

Review It

Life depends on a constant input of energy to maintain the structure of cell. Identify the two organelles that specialize in converting energy to a usable form.

Define what a plastid is, and list three that can be found in plant cells.

Use It

Chloroplasts have a third membrane called the thylakoids. Describe what role they play.

Mitochondria have two membranes. How is the inner membrane different than the outer membrane?

Describe two important features about chloroplasts and the mitochondria which support the endosymbiotic theory.

4.8 The Cytoskeleton

Essential Knowledge covered
1.B.1: Organisms share many conserved core processes and features that evolved and are widely distributed among organisms today.
2.B.3: Eukaryotic cells maintain internal membranes that partition the cell into specialized regions.
4.A.2: The structure and function of subcellular components, and their interactions, provide essential cellular processes.
4.B.2: Cooperative interactions within organisms promote efficiency in the use of energy and matter.

Recall It

The cytoskeleton is a dynamic network of proteins, giving the cell shape and allowing the cell and its organelles to move. Proteins that attach, detach, and reattach to filaments in the cell are called motor molecules. There are three types of filaments. Actin filaments are long, very thin, and made up two chains of twisted globular actin monomers. Between the size of actin filaments and microtubules, are the intermediate filaments that form ropelike assemblies of polypeptides. Microtubules are made of globular tubulin and controlled through the centrosome. A short cylinder of microtubules arranged in an outer ring is called a centriole. The organelle called the basal body may direct the organization of microtubules within cilia and flagella.

Review It

Compare and contrast the function of actin filaments and microtubules in a cell.

4.8 The Cytoskeleton *continued*

Identify the fiber of the cytoskeleton based on fiber's composition.

Composition	Fiber
Globular proteins of actin twisted together	
Tetramers of protein vimentin	
α- and β-tubulin protein subunits arranged side-by-side	

Use It

Determine what cytoskeletal fiber is responsible for the following action or trait:

Action/Trait	Fiber
An animal cell pinches into two cells during cellular division	
A chimpanzee has hard fingernails	
A sparrow flaps its wing	
A vesicle is transported along the ER	
Chromosomes are pulled apart during mitosis	
A human has hard fingernails	

In your own words, how does a chameleon use the cytoskeleton to change the color of its skin to blend into its environment?

Describe how cilium and flagella are able to move using the cytoskeleton.

AP CHAPTER SUMMARY

Summarize It

What is the central dogma of molecular biology, and how does it govern all living cells?

Explain how a ribosome synthesizing a protein with a signal peptide moves from the nucleus to the endoplasmic reticulum.

In her 1967 paper, Dr. Lynn Margulis described the how eukaryotic cells arose from prokaryotic cells through her endosymbiotic theory.

Using the diagram below, label and describe how organelles are thought to have evolved, explaining how the internal membranes around organelles support her theory.

Prokaryotic cells   Eukaryotic Cell

While only plants and algae contain chloroplasts, all eukaryotes contain mitochondria which produce ATP. Why is ATP important?

Choose one of the following diagrams to answer the questions below:

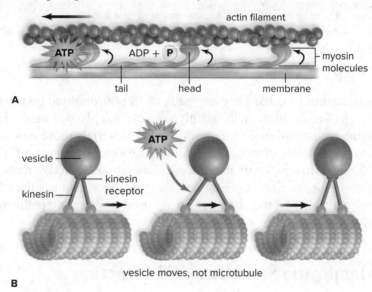

Identify the filament and motor protein.

Describe the interaction that is occurring.

What process in a living organism may be the result of this interaction?

5 Membrane Structure and Function

FOLLOWING *the* BIG IDEAS

	AP Essential Knowledge	Chapter Section
BIG IDEA 2	**2.B.1** Cell membranes are selectively permeable due to their structure.	5.2, 5.4
	2.B.2 Growth and dynamic homeostasis are maintained by the constant movement of molecules across membranes.	5.1, 5.2, 5.3
	2.C.2 Organisms respond to changes in their external environments.	5.4
BIG IDEA 3	**3.A.1** DNA, and, in some cases RNA, is the primary source of heritable information.	5.1
	3.B.2 A variety of intercellular and intracellular signal transmissions mediate gene expression.	5.1
	3.D.1 Cell communication processes share common features that reflect a shared evolutionary history.	5.1
	3.D.2 Cells communicate with each other through direct contact with other cells or from a distance via chemical signaling.	5.4
BIG IDEA 4	**4.C.1** Variation in molecular units provides cells with a wider range of functions.	5.1

CHAPTER OVERVIEW

Membranes are key structures of cells. They are made up of phospholipid bilayers and can contain a number of other types of molecules, including glycolipids, and glycoproteins. The fluid-mosaic model describes the interactions of molecules within cells. Materials move across membranes in a number of different ways. Passive transport does not require energy as molecules move from high to low concentrations via diffusion, facilitated diffusion, or osmosis. Active transport requires the use of energy when molecules need to move against their gradients and use the help transport proteins. Proteins embedded in cell membranes also are used in signaling pathways for cell communication.

5.1 Plasma Membrane Structure and Function

Essential Knowledge covered
2.B.1: Cell membranes are selectively permeable due to their structure.
2.B.2: Growth and dynamic homeostasis are maintained by the constant movement of molecules across membranes.
3.B.2: A variety of intercellular and intracellular signal transmission mediate gene expression.
3.D.1: Cell communication processes share common features that reflect a shared evolutionary history.
3.D.2: Cells communicate with each other through direct contact with other cells or from a distance via chemical signaling.
4.C.1: Variation in molecular units provides cells with a wide range of functions.

Recall It

Cell membranes contain many different proteins and phospholipids, cholesterol, glycoproteins, and glycolipids. The fluid-mosaic model describes how the interactions of membrane components interact, allowing the membrane to be flexible and perform many different functions. Some proteins are embedded in the membrane, while some span across the bilayer, and some are bound to the surface. Phospholipids give membranes amphipathic properties, with hydrophilic and hydrophobic regions of the lipid bilayer. Proteins associated with cell membranes provide six major functions: 1) transport of polar or large molecules; 2) enzymatic activity; 3) signal transduction; 4) cell-cell recognition; 5) intercellular joining; and 6) attachment of the cytoskeleton to the extracellular matrix. Cell membranes are selectively permeable.

5.1 Plasma Membrane Structure and Function *continued*

Review It

Use your textbook to identify the structure or function associated with the plasma membrane:

Definition	Structure or Function
Only allows certain substances in, while keeping others out	
Channel proteins that allow water across a membrane	
Where molecules flow from where their concentration is high to where their concentration is low	
A phospholipid attached to a carbohydrate chain	
A way in which large particles can exit or enter a cell	
Protein and carbohydrate molecules found outside of animal cells	
A protein with an attached carbohydrate chain	

Use It

Describe what causes cystic fibrosis and how it affects the body.

Why do cells respond only to certain signaling molecules?

5.1 Plasma Membrane Structure and Function *continued*

Identify each membrane protein and describe its function.

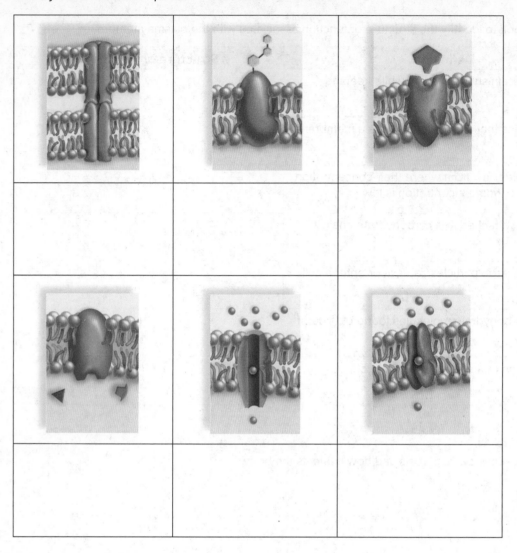

Carbon dioxide is an example of a small, noncharged molecule that can pass across a cell's membrane. Use an arrow to indicate which way the CO_2 will move through this membrane.

CO_2 CO_2 CO_2 CO_2 CO_2 CO_2		CO_2 CO_2 CO_2
outside cell	cell membrane	inside cell

5.2 Passive Transport Across a Membrane

Essential Knowledge covered
2.B.1: Cell membranes are selectively permeable due to their structure.
2.B.2: Growth and dynamic homeostasis are maintained by the constant movement of molecules across membranes.

Recall It

Passive transport does not require the input of energy to move molecules across membranes. Diffusion, osmosis, and facilitated diffusion are three different forms of passive transport found in membranes. Diffusion is the movement of molecules from a higher to a lower concentration, whereas osmosis is the diffusion of water. Osmotic pressure develops as a result of osmosis. Remember that solutions contain both solutes and solvents. In isotonic solutions, the solute concentration and water concentration both inside and out are equal, so there is no movement of water. Cells placed in hypotonic solutions swell because these solutions have a lower concentration of solute than the cell. Hypertonic solutions cause cells to shrink because they have a higher concentration of solute than the cell. Facilitated transport moves molecules such as glucose and amino acids using the aid of carrier proteins. Water can move across membranes in a process called osmosis and flows through specialized channels for water called aquaporins.

Review It

Compare and contrast diffusion and facilitated transport.

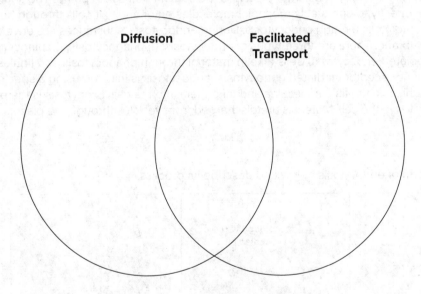

Use It

A laboratory technician performed a routine red blood cell count. The cells appeared shriveled and were difficult to count. What was wrong with the solution the cells were stored in?

The technician from the question above then noticed that the cells were not in the correct solution and diluted the solution to a concentration of 0.5% sodium chloride. What happened to the cells?

5.2 Passive Transport Across a Membrane *continued*

Imagine that you find a wilted piece of celery lurking in your refrigerator. You take it out and put it in a glass of water. The next day, the celery stalk is firm again. Explain what has happened at the cellular level. Use the words *vacuole, osmotic pressure, turgor pressure, hypertonic,* and *hypotonic* in your answer.

5.3 Active Transport Across a Membrane

Essential Knowledge covered
2.B.2: Growth and dynamic homeostasis are maintained by the constant movement of molecules across membranes.

Recall It

Active transport requires the use of energy to move molecules against their concentration gradient. Like facilitated diffusion, active transport requires specialized carrier proteins to move certain materials into or out of the cell. An example of a carrier protein is the sodium-potassium pump. The sodium-potassium pump uses ATP directly to operate. Large, polar molecules enter and exit cells through bulk transport. Bulk materials enter cells through endocytosis. During endocytosis, substances are enveloped by the plasma membrane. There are three types of endocytosis: (1) phagocytosis, (2) pinocytosis, and (3) receptor-mediated endocytosis. Cells take in material through phagocytosis and liquids through pinocytosis. During receptor-mediated endocytosis, endocytosis is initiated using a specific receptor protein to recognize compatible molecules and take them into the cell. Exocytosis is the process in which materials leave the cell. Materials are discharged from vesicles through the plasma membrane.

Review It

Identify each form of endocytosis below and describe its process.

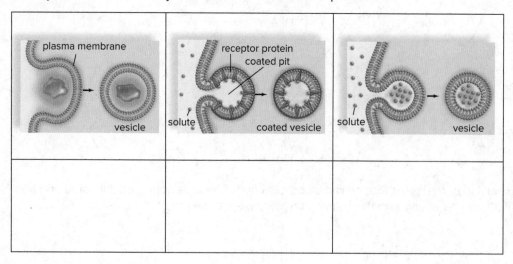

5.3 Active Transport Across a Membrane *continued*

Use It

Design a flowchart to illustrate how a cell secretes substances outside of the cell through exocytosis. Be sure to label the plasma membrane, the secretory vesicle, and the substance being transferred.

5.4 Modification of Cell Surfaces

Essential Knowledge covered
2.B.1: Cell membranes are selectively permeable due to their structure.
2.C.2: Organisms respond to changes in their external environment.
3.D.2: Cells communicate with each other through direct contact with other cells or from a distance via chemical signaling.

Recall It

The extracellular matrix (ECM) outside of animal cells aids in support, structure, and communication. There are many types of molecules that give the ECM its properties, such as collagen, elastin fibers, fibronecin, and intergrins. The ECM influences the activities of the cytoskeleton of the cell. There are many types of junctions that exist between cells, including (1) adhesion junctions, (2) tight junctions, (3) gap junctions, and (4) plasmodesmenta which are found in plant cells.

Review It

There are many components in the extracellular matrix. This section briefly covered some examples. Describe what they were and how they function.

The Extracellular Matrix	
Components	**Function**
Collagen and elastin	
	Resists compression and assist in cell signaling
	Influences the shape and activity of the cell by allowing for communication between the EMC and cytoskeleton

Place an X under the junction where the following statement applies.

	Desmosomes	**Tight Junction**	**Gap Junction**
This junction allows communication between two cells by joining plasma membrane channels			
This junction forms an impermeable barrier between adjacent cells			
Mechanically attaches to adjacent cells			

Use It

Identify the junction that helps play a role in the following organismal processing.

Process	Junction
Keeps digestive juice only moving between intestinal cells	
Passes water from plant cell to plant cell	
Allows for a flow of ions to between heart cells, permitting contractions	
Holds skin cells together and allows skin to remain elastic	

All plants have cell walls. Describe the importance of these three components to a cell wall and where they are found.

Pectin	Plasmodesmata	Lignin

Why might plants have evolved cell walls while animals did not?

AP CHAPTER SUMMARY

Summarize It

How does the fluid mosaic model of the cell membrane allow for selective permeability?

Organisms are able to grow, sense, and respond to their environment due to specific signals that occur in cells. How do cells talk to one another?

How do membrane-bound organelles in eukaryotic cells confer greater efficiency to cell processes?

FOLLOWING *the* BIG IDEAS

AP Essential Knowledge	Chapter Section
BIG IDEA 2 **2.A.1** All living systems require constant input of free energy.	6.1, 6.2, 6.3, 6.4
2.A.2 Organisms capture and store free energy for use in biological processes.	6.1, 6.2, 6.4
2.D.1 All biological systems from cells and organisms to populations, communities and ecosystems are affected by complex biotic and abiotic interactions involving exchange of matter and free energy.	6.3
BIG IDEA 3 **3.A.1** DNA, and, in some cases RNA, is the primary source of heritable information.	6.3
BIG IDEA 4 **4.A.2** The structure and function of subcellular components, and their interactions, provide essential cellular processes.	6.3
4.B.1 Interactions between molecules affect their structure and function.	6.3

CHAPTER OVERVIEW

All organisms need energy in order to maintain structure and to perform metabolic activities. Energy used by organisms on Earth comes from the energy in sunlight. This solar energy is converted by plants through photosynthesis into organic nutrients. The behavior and transfer, or flow, of energy is described by the field of thermodynamics. The metabolism of energy is described through chemical reactions. ATP is known as the energy currency of the cell. Enzymes play an important role in chemical reactions.

6.1   Cells and the Flow of Energy

Essential Knowledge covered
2.A.1: All living systems require constant input of free energy.
2.A.2: Organisms capture and store free energy for use in biological processes.

Recall It

Energy can come in two forms: kinetic energy and potential energy. Kinetic energy is the energy of motion, whereas potential energy is energy that is stored. Energy mainly enters the biological world through sunlight, where it is converted through photosynthesis into sugars. There are two laws of thermodynamics discussed in this chapter. The first law states that energy cannot be created or destroyed; it can only change from one form to another. The second law states that entropy, the disorder in the universe, is continuously increasing. No process requiring a conversion of energy is ever 100% efficient, so the amount of energy something has is commonly measured as the loss of heat in biology.

Review It

Provide a definition of energy.

List the two forms in which energy occurs.

6.1 Cells and the Flow of Energy *continued*

What are the two laws of thermodynamics?

Define *entropy*.

Use It

Identify the form of energy in each description as either kinetic (K) or potential (P).

Energy Form	Description
	the monkey swinging through a tree
	the fig that the monkey eats
	molecules in the fig that the monkey eats
	the monkey eating a fig

Describe how solar energy activates the light phases in a plant.

Use the second law of thermodynamics to describe why and how glucose breaks down over time.

Compare and contrast chemical energy and mechanical energy.

6.2 Metabolic Reactions and Energy Transformations

Essential Knowledge covered
2.A.1: All living systems require constant input of free energy.
2.A.2: Organisms capture and store free energy for use in biological processes.

Recall It

The sum of all chemical reactions that occur in a cell is called metabolism. The amount of energy available to do work after a chemical reaction has occurred is free energy. Spontaneous reactions are called exergonic, while reactions that require an input of energy are called endergonic. The most common energy currency of cells is ATP. ATP can be regenerated from ADP and inorganic phosphate. The hydrolysis of ATP to ADP and inorganic phosphate can be measured in kcal per mole.

Review It

Identify the reactants and products in this equation.
A + B → C + D

List three ways which ATP provides energy for living organisms.

Use It

Illustrate the ATP cycle in terms of the creation and hydrolysis of ATP. Identify which reaction is endergonic and which is exergonic.

What does it mean when a product has been *phosphorylated*? You may find it helpful to draw a picture of the coupling of ATP to an energy-requiring reaction in order to help you answer this question.

In your own words, describe how ATP is used to contract a muscle which is illustrated in the figure below.

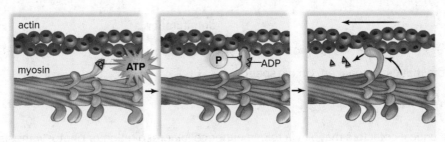

6.3 Metabolic Pathways and Enzymes

Essential Knowledge covered

2.A.1: All living systems require constant input of free energy.

2.D.1: All biological systems from cells and organisms to populations, communities and ecosystems are affected by complex biotic and abiotic interactions involving exchange of matter and free energy.

3.A.1: DNA, and in some cases RNA, is the primary source of heritable information.

4.A.2: The structure and function of subcellular components, and their interactions, provide essential cellular processes.

4.B.1: Interactions between molecules affect their structure and function.

6.3 Metabolic Pathways and Enzymes *continued*

Recall It

Chemical reactions between molecules involve both bond breaking and bond forming. When the bonds form, energy is released as heat. To start a reaction, the substrates or reactants must absorb energy to reach an unstable state where the bonds can break. This energy is known as the energy of activation. An enzyme is a biological catalyst to regulate chemical reactions. Catalysts lower the activation energy needed for a reaction to occur by lowering the activation energy barrier. An enzyme might do this by bringing two substrates together in the correct orientation or by stressing particular chemical bonds. There are many different types of enzymes, each which catalyze specific chemical reactions. Enzymes are sensitive to environmental factors, including pH and temperature, as well as the presence of regulatory molecules.

Review It

Insert the correct definition or vocabulary word in the list below.

Definition	Vocabulary Word
a series of linked reactions	
	active site
a protein molecule that speeds up a chemical reaction	
	ribozymes
a frequent component of a coenzyme often found in our diets	
	co-factors/co-enzymes
reactants in enzymatic reactions	
	denaturation

List four factors that can impact enzymatic speed.

List the three critical cofactors that play a role in either cellular respiration or photosynthesis.

Use It

Using the enzyme and substrate below, describe the induced fit model.

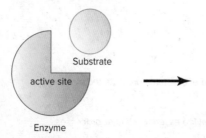

The diagram below illustrates a simple metabolic pathway, where E stands for the enzyme needed in order to complete a reaction:

$$A \xrightarrow{E_1} B \xrightarrow{E_2} C$$

If E_1 became denatured, what would happen to the pathway?

What do **A** and **B** represent? How do they differ from one another?

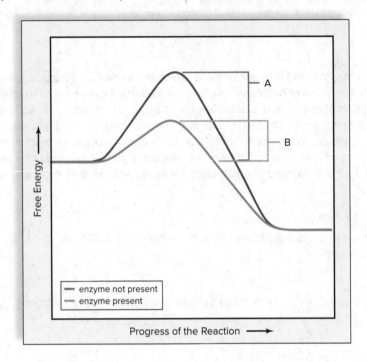

Free Energy →

Progress of the Reaction →

— enzyme not present
— enzyme present

If the enzyme in the reaction above needed to be in solution with a pH = 5 in order to maintain structural integrity, and the solution was at pH 8, what would happen to the reaction?

Decide whether or not the following factors affect the rate of an enzymatic reaction. Place a + if the factor increases the rate, a − if it decreases the rate, or an = sign if the rate stays the same.

Factor	Rate of Reaction
The concentration of substrate increases	
An inhibitor is present	
Optimal pH has been obtained	
A cofactor is missing	

6.3 Metabolic Pathways and Enzymes *continued*

Some animals, such as Siamese cats, have what is called point coloration where the warmest parts of the body are pale and the cooler extremities are darker in color. This is caused by a mutated gene that regulates an enzyme involved in melanin production. How might this enzyme work?

6.4 Oxidation-Reduction Reactions and Metabolism

Essential Knowledge covered
2.A.1: All living systems require constant input of free energy.
2.A.2: Organisms capture and store free energy for use in biological processes.

Recall It

Reactions that involve the gain and loss of electrons are called oxidation-reduction reactions. The process of photosynthesis and the process of cellular respiration are important examples of oxidation-reduction reactions. Chloroplasts in plants capture solar energy and use it to convert water and carbon dioxide into a carbohydrate. Mitochondria, present in both plants and animals, oxidize carbohydrates and use the released energy to build ATP molecules in a process called cellular respiration. Cellular respiration consumes oxygen and produces carbon dioxide and water. These organelles are involved in a redox cycle, because carbon dioxide is reduced during photosynthesis and carbohydrate is oxidized during cellular respiration.

Review It

Describe what happens in a redox reaction (think of the term OIL RIG).

List the two important metabolic pathways to all of life that contain redox reactions.

Name the two organelles involved in a redox cycle.

Satisfy the following equations.

A. energy + _____ CO_2 + 6_____ → _____ + 6 O_2

B. $C_6H_{12}O_6$ + 6 _____ → _____ + $6CO_2$ + $6H_2O$

Identify which equation above describes cellular respiration and which describes photosynthesis.

Use It

Which molecules are reduced and which are oxidized in photosynthesis?

Which molecules are reduced and which are oxidized in cellular respiration?

Illustrate the redox cycle that occurs between a mitochondria and a chloroplast. Be sure to include solar energy, ATP, heat, and the molecules involved in the redox reaction.

AP CHAPTER SUMMARY

Summarize It

The human body produces a lot of heat. Describe this heat in terms of solar energy and entropy.

How does the structure of ATP enable the molecule to power cellular work?

Compare and contrast these two different forms of enzyme inhibition: noncompetitive inhibition and competitive inhibition.

FOLLOWING *the* BIG IDEAS

AP Essential Knowledge	Chapter Section
BIG IDEA 2 **2.A.1** All living systems require constant input of free energy.	7.2, 7.3, 7.4, 7.5
2.A.2 Organisms capture and store free energy for use in biological processes.	7.1, 7.2, 7.3, 7.4, 7.5
BIG IDEA 4 **4.C.1** Variation in molecular units provides cells with a wider range of functions.	7.3

CHAPTER OVERVIEW

Photosynthesis is the process by which some organisms can capture energy from sunlight and convert it into chemical energy. Photosynthesis can be described in a series of biochemical reactions and cycles; the light cycle and Calvin cycle. The process of photosynthesis evolved in bacteria, algae, and plants, and allowed for the diversity of life as we know it on Earth. The oxygen we breathe, and even the energy we extract from food molecules, all depend on the process of photosynthesis.

7.1 Photosynthetic Organisms

Essential Knowledge covered
2.A.2: Organisms capture and store free energy for use in biological processes.

Recall It

Autotrophs produce their own food through photosynthesis; the process of converting solar energy into chemical energy, stored as carbohydrates. Photosynthetic organisms, or producers, include plants, algae, and cyanobacteria. Photosynthesis takes place in specialized cells in organelles called chloroplasts. Chloroplasts have thylakoid membranes, and the interior of the chloroplast is called the stroma. Consumers who depend on the energy that autotrophs create are called heterotrophs.

Review It

Describe and differentiate a heterotroph and an autotroph.

Use It

Identify if the organism is a heterotroph (H) or an autotroph (A).

_____ macroinverebrate

_____ plant

_____ mammal

_____ cyanobacteria

_____ algae

7.2 The Process of Photosynthesis

Essential Knowledge covered
2.A.1:   *All living systems require constant input of free energy.*
2.A.2:   *Organisms capture and store free energy for use in biological processes.*

Recall It

Photosynthesis is the conversion of solar energy into chemical energy. Photosynthesis combines CO_2 and H_2O to produce glucose and O_2. During the light reactions, solar energy is used to synthesize ATP and NADPH. Then, in the Calvin cycle, ATP and NADPH are used to convert CO_2 into carbohydrates. The light reactions take place in the thylakoid membranes of the chloroplasts, and the Calvin cycle occurs in the stroma.

Review It

Reduction and oxidation take place in photosynthesis.
Label the diagram below to show which molecules are reduced and which are oxidized

$$solar\ energy$$
$$CO_2 + H_2O \longrightarrow (CH_2O) + O_2$$

Identify which equation represents the light reaction and which equation represents the Calvin cycle.

The Two Stages of Photosynthesis	
solar energy → chemical energy (ATP, NADPH)	chemical energy → chemical energy (ATP, NADPH) (carbohydrate)

Use It

Explain the role of $NADP^+$/NADPH in photosynthesis and how each chemical is connected to the process.

Label the light reaction and Calvin reaction and the energy that powers the reactions.

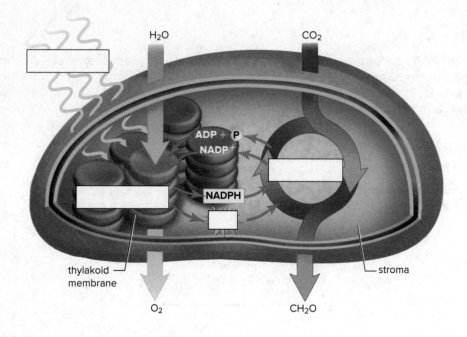

Compare and contrast the light reactions and Calvin cycle.

7.3 Plants Convert Solar Energy

Essential Knowledge covered
2.A.1: *All living systems require constant input of free energy.*
2.A.2: *Organisms capture and store free energy for use in biological processes.*
4.C.1: *Variation in molecular units provides cells with a wide range of functions.*

Recall It

Pigments capture solar energy for photosynthesis. Chlorophylls *a* and *b* are pigments that play important roles in photosynthesis, and carotenoids are accessory pigments. Pigment molecules become "antennas" in plant photosystems, which gather energy from photons and then excite electrons to higher energy levels in locations called reaction centers. Light reactions use photosystem II (PS II) and photosystem I (PS I) to perform photosynthesis in the noncyclic pathway. PS II receives electrons from water as water splits and releases oxygen, and PS I reduces $NADP^+$ to NADPH. PS I and PS II are connected by the transfer of higher free energy electrons through an electron transport chain (ETC), which pumps H^+ across the thylakoid membrane by chemiosmosis. Chemiosmosis is a gradient used to make ATP via ATP synthase. Some bacteria use a cyclic pathway, in which they generate ATP but not NADPH using PS I.

Review It

List four molecular complexes that make up the thylakoid membrane.

Use It

Why do plants appear green to us?

Illustrate and describe how ATP and NADPH are used in PS II and PS I.

7.4 Plants Fix Carbon Dioxide

Essential Knowledge covered
2.A.1: All living systems require constant input of free energy.
2.A.2: Organisms capture and store free energy for use in biological processes.

Recall It

The Calvin cycle reactions use the energy captured in the light reactions for carbon-fixation. The Calvin cycle uses CO_2, ATP, and NADPH to build simple sugars. The Calvin cycle occurs in three stages: (1) the enzyme RuBP carboxylase "fixes" carbon from CO_2 to RuBP, which produces an unstable six-carbon molecule that spills into two 3-carbon molecules, (2) the reduction of the resulting 3-carbon PGA to G3P, generating ATP and NADPH, and (3) the regeneration of RuBP. Six turns of the cycle fix enough carbon to produce two excess G3Ps. It takes two G3P molecules to produce one molecule of glucose ($C_6H_{12}O_6$).

Review It

What are the major products of the Calvin cycle and why are they important?

List the major steps of the Calvin cycle.

7.4 Plants Fix Carbon Dioxide *continued*

Use It

Describe two steps which make G3P. Be sure to explain how ATP and NADPH are used.

Identify two possible fates of G3P.

Using the diagram, identify the metabolites of the Calvin cycle with an arrow. Fill in the spaces with the correct molecules used to drive the different steps of the Calvin cycle.

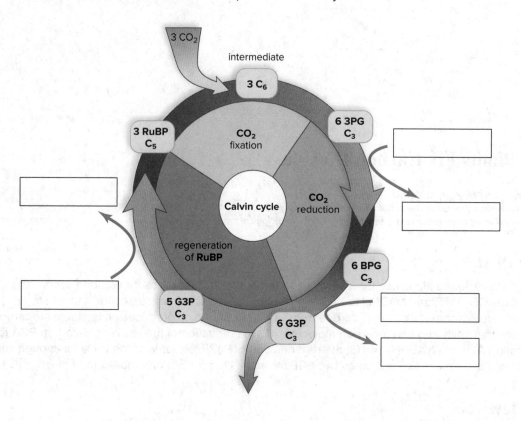

Where were these molecules that provide the energy and electrons for the reduction reactions produced?

How many G3P molecules are needed to form glucose?

7.5 Other Types of Photosynthesis

Essential Knowledge covered
2.A.1: *All living systems require constant input of free energy.*
2.A.2: *Organisms capture and store free energy for use in biological processes.*

Recall It

The majority of plants are C_3 plants and carry out photosynthesis using RuBP carboxylase to fix CO_2 to RuBP. Photorespiration occurs when stomata close to conserve water, causing oxygen to join with RuBP instead of CO_2. In order to stop this wasteful reaction, C_4 plants adapted to fix CO_2 to PEP with the enzyme PEPCase. Plants that perform CAM photosynthesis fix CO_2 at night and release C_4 to the Calvin cycle during the day.

Review It

Complete the following chart of photosynthesis types and how they fix CO_2.

Photosynthesis Type	CO_2 Fixing Enzyme	Cells Involved
	RuBP	
CAM		Mesophyll cells and bundle sheath cells
		Mesophyll cells

Identify the following processes.

$$RuBP + CO_2 \xrightarrow{\text{RuBP carboxylase}} 2\ 3PG \qquad \underline{\hspace{3cm}}$$

$$PEP + CO_2 \xrightarrow{\text{PEPCase}} \text{oxaloacetate} \qquad \underline{\hspace{3cm}}$$

$$RuBP + O_2 \xrightarrow{\text{RuBP carboxylase}} 3PG + CO_2 \qquad \underline{\hspace{3cm}}$$

Use It

Determine whether or not the following statements are true or false (T/F) about the different types of photosynthesis.

Plants with CAM photosynthesis are most likely found in environments where temperatures are below 25°C.

C_4 plants partition the fixation of CO_2 temporally.

C_3 photosynthesis partitions CO_2 in mesophyll cells and the Calvin cycle in bundle sheath cells.

C_4 plants partition the fixation of CO_2 in different spaces.

What is the primary advantage for partitioning photosynthesis temporally in CAM plants?

Summarize It

How do plants capture solar energy and convert it to chemical energy?

Where do electrons come from that are used in PS II, and how are they transferred to PS I?

When chlorophyll breaks down in leaves during the fall, why do they appear yellow and orange?

In your own words, describe how C4 and CAM plants are able to live in hot, dry conditions and why C_3 plants have problems.

FOLLOWING *the* BIG IDEAS

AP Essential Knowledge	Chapter Section
BIG IDEA 2 **2.A.1** All living systems require constant input of free energy.	8.1, 8.2, 8.3, 8.4, 8.5
2.A.2 Organisms capture and store free energy for use in biological processes.	8.1, 8.2, 8.3, 8.4, 8.5

CHAPTER OVERVIEW

All organisms carry out some form of cellular respiration. Cellular respiration is an oxidation-reduction reaction that allows cells to acquire energy. Cellular respiration can be summarized by the following equation:

$$C_6H_{12}O_6 + 6O_2 \rightarrow 6CO_2 + 6H_2O + \text{energy}$$

The complete breakdown of glucose occurs in four phases, and yields ATP – the energy that all cells need in order to process.

8.1 Overview of Cellular Respiration

Essential Knowledge covered
2.A.1: All living systems require constant input of free energy.
2.A.2: Organisms capture and store free energy for use in biological processes.

Recall It

Cellular respiration is the process by which cells acquire energy. Cellular respiration involves the breakdown of glucose into carbon dioxide and water. There are four phases of cellular respiration: (1) glycolysis, (2) the prep reaction, (3) the citric acid cycle, and (4) the electron transport chain. NAD^+ and FAD are two coenzymes that play critical roles in the oxidation-reduction reactions that occur during cellular respiration.

Review It

List the four phases of the complete breakdown of glucose. Identify whether each phase occurs inside or outside the mitochondria, and whether the phases are anaerobic or aerobic.

8.1 Overview of Cellular Respiration *continued*

Fill in the diagram below to show the order of the four phases of cellular respiration. In each box, place the name of the phase and the main products produced. Place a star by the cycles that turn twice.

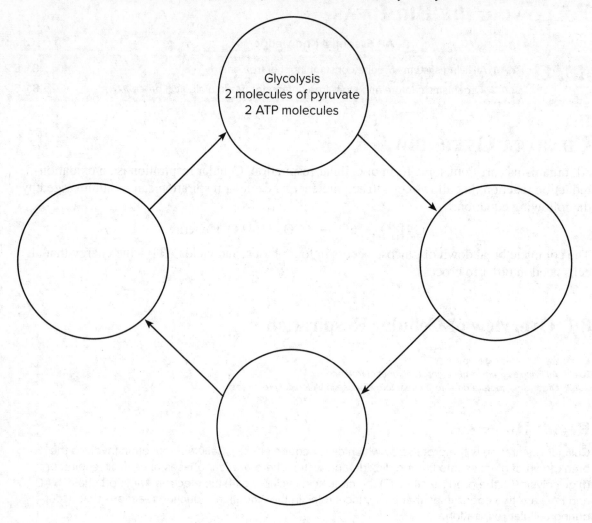

Use It

Explain why the breakdown of glucose happens in steps.

What are the main functions of NAD$^+$ and FAD in cellular respiration?

8.2 Outside the Mitochondria: Glycolysis

Essential Knowledge covered
2.A.1: All living systems require constant input of free energy.
2.A.2: Organisms capture and store free energy for use in biological processes.

Recall It

Glycolysis is the breakdown of C_6 carbon glucose molecules to two C_3 pyruvate molecules. Glycolysis is an anaerobic reaction, and likely evolved before the citric acid cycle. Glycolysis occurs in a series of ten steps. Pathways are divided into energy-investment steps and energy-harvesting steps. During glycolysis, there is a net gain of two ATP molecules. ATP synthase is formed through substrate-level phosphorylation. When O_2 is available, the end product of glycolysis, pyruvate, enters the mitochondria, where it is metabolized. If O_2 is not available, fermentation occurs.

Review It

The formation of 36 to 38 ATP are theoretically possible when glucose is broken down completely. How many ATP molecules are produced during glycolysis?

Identify the inputs and outputs of glycolysis.

Glycolysis	
Inputs	**Outputs**
6C glucose	
2 ATP	2 NADH

Where does glycolysis occur?

Use It

Substrate-level phosphorylation occurs in the later steps of glycolysis. Illustrate and describe how an enzyme might pass phosphate to ADP to form ATP. Be sure to label the enzyme and molecules.

8.3 Outside the Mitochondria: Fermentation

Essential Knowledge covered
2.A.1: All living systems require constant input of free energy.
2.A.2: Organisms capture and store free energy for use in biological processes.

Recall It

Fermentation is an anaerobic process that produces a limited amount of ATP in the absence of oxygen. It occurs outside of the mitochondria. Fermentation consists of glycolysis followed by a reduction of pyruvate. NAD^+ is recycled and returned to the glycolytic pathway to pick up more electrons. Each step is catalyzed by a specialized enzyme. Depending on their particular enzymes, bacteria vary as to whether they produce an organic acid, such as lactate, or an alcohol and CO_2. In animal cells, lactate is always the end product of glycolysis.

Review It

The formation of 36 to 38 ATP are theoretically possible when glucose is broken down completely. How many ATP molecules are produced during fermentation?

Identify the inputs and outputs of fermentation.

Fermentation	
Inputs	**Outputs**
glucose	2 lactate or 2 alcohol

Where does fermentation occur?

Use It

Identify the type of fermentation possible by each organism (**L** for lactate acid fermentation or **A** for those that produce ethyl alcohol as a result).

Animals

Plants

Yeast

Bacteria

Compare and contrast the two forms of fermentation.

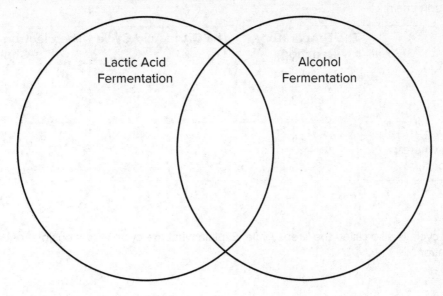

Lactic Acid Fermentation

Alcohol Fermentation

8.4 Inside the Mitochondria

Essential Knowledge covered
2.A.1: All living systems require constant input of free energy.
2.A.2: Organisms capture and store free energy for use in biological processes.

Recall It

The preparatory (prep) reaction, the citric acid cycle, and the electron transport chain are needed for the complete breakdown of glucose. These reactions take place inside the mitochondria. The prep cycle converts C_3 pyruvate into a C_2 acetyl and releases CO_2. The C_2 acetyl is used in the citric acid cycle, where more CO_2 is released, and ATP is synthesized from ADP and inorganic phosphate by substrate-level phosphorylation. Electrons extracted from the intermediate organic molecules are carried by NADH and $FADH_2$ to the electron transport chain. Here, electrons delivered by NADH and $FADH_2$ are passed to a series of electron acceptors embedded in the membrane as they move toward the final electron acceptor, oxygen. The flow of hydrogen ions creates a proton gradient, and powers ATP synthase by chemiosmosis. This generates ATP from ADP and inorganic phosphate.

Review It

Explain the reactions that occur in the mitochondria and why the organelle is often referred to as the powerhouse of the cell.

Fill in the following sentences with the correct word or number.

- A maximum of _____ ATP molecules may be produced by the electron transport chain.
- The _____ converts products from glycolysis into products that enter the citric acid cycle.
- It takes _____ turns of the citric acid cycle to process each original glucose molecule.
- Electrons that enter the electron transport chain are carried by _____.

8.4 Inside the Mitochondria *continued*

Fill in the following information concerning the preparatory reaction, the citric acid cycle, and the electron transport chain.

	The Preparatory Reaction	The Citric Acid Cycle	The Electron Transport Chain
Products			
Location in mitochondria where the process occurs			

Use It

The citric acid cycle is also called the Krebs cycle. Explain what the cycle is responsible for during cellular respiration.

What is the primary function of oxygen in cellular respiration?

If oxygen was not present during cellular respiration, what would happen?

8.5 Metabolism

Essential Knowledge covered
2.A.1: *All living systems require constant input of free energy.*
2.A.2: *Organisms capture and store free energy for use in biological processes.*

Recall It

Molecules needed for metabolic pathways are said to be stored in metabolic pools. Metabolic pools are added to through the constructive reactions called anabolism and broken down through catabolism. Proteins are sometimes also use as an energy source in which amino acids may undergo a process called deamination, thereby losing their amino group.

Review It

Name the two organelles instrumental in allowing the flow of energy through living organisms.

8.5 Metabolism *continued*

Mitochondria and chloroplasts perform opposite processes but have very similar structures in their organelles. Using the table below, describe the differences in the organelles.

Structure	Chloroplast	Mitochondria
Inner membrane		
Electron Transport Chain		
Enzymes		

Use It

Determine whether or not the following statements are true or false (T/F) about the metabolic pool.

The balance of catabolism and anabolism is essential for optimum cellular function.

Anabolism is the breakdown of molecules.

Catabolism is the building of new molecules.

Products from catabolic processes are needed in order to build new molecules.

Why is fat an efficient form of stored energy?

AP CHAPTER SUMMARY

Summarize It

Describe how the human body produces ATP after a person eats a large bowl of pasta.

Cyanide is a poisonous substance because it binds to a redox carrier called cytochrome. Why would this be dangerous?

Describe how energy flows from the sun to a chloroplast in a plant and then to a mitochondria in a human. Feel free to illustrate your answer.

UNIT ONE: THE CELL

Chapter 2 Basic Chemistry • Chapter 3 The Chemistry of Organic Molecules • Chapter 4 Cell Structure and Function • Chapter 5 Membrane Structure and Function • Chapter 6 Metabolism: Energy and Enzymes • Chapter 7 Photosynthesis • Chapter 8 Cellular Respiration

Multiple Choice Questions

Directions: For each question or incomplete statement chose one of the four suggested answers or completions listed below.

1. A muscle cell with a sodium-potassium pump carries three sodium ions outward for every two potassium ions carried inward. Which statement is true regarding the state of the cell?

   (A) The outside of the cell carries less of a positive charge compared to the inside of the cell.

   (B) The inside of the cell carries less of a positive charge compared to the outside of the cell.

   (C) The state of the cell depends on if the muscle is being exercised.

   (D) The cell is in equilibrium.

2. Why is glycolysis thought to have evolved before the citric acid cycle and the electron transport chain?

   (A) Glycolysis occurs before the citric acid cycle.

   (B) Glycolysis breaks down C_6 glucose to two C_3 pyruvate molecules

   (C) Glycolysis occurs universally in all organisms and does not require oxygen.

   (D) Glycolysis uses substrate-level ATP synthesis to form ATP.

3. White spot syndrome virus (WSSV) is lethal to Penaeid shrimp species. Scientists performed an experiment to identify how WSSV binds to and affects Penaeid shrimp cells. It was hypothesized that BP53, a subunit of ATP-synthase, present in shrimp cells, acts as a receptor for WSSV.

The following figure shows the data from four experimental groups.

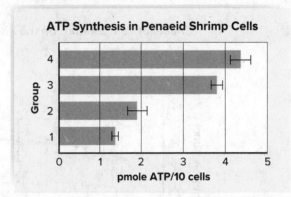

TREATMENT KEY

Group 1: WSSV infected cells incubated with anti-rBP53 antibody
Group 2: WSSV naturally infected shrimp cells
Group 3: WSSV bound to the cell surfaces
Group 4: Control

Liang, Yan, et al. "ATP synthesis is active on the cell surface of the shrimp Litopenaeus vannamei and is suppressed by WSSV infection." *Virology journal* 12.1 (2015): 49.

Which of the following statements is NOT supported by these data?

(A) ATP synthesis may occur on the surface of shrimp cells.

(B) WSSV suppresses ATP production.

(C) BP53 suppresses ATP production.

(D) rBP53 antibody with WSSV infected cells further suppresses ATP production.

4. Scientists evaluated how a diet high in sugar and starch compared with a diet high in fat and fiber affected glucose utilization in trained Arabian horses during exercise. It was found that horses fed a diet higher in sugar and starch had higher circulating glucose but a higher glucose irreversible loss during exercise. With these results in mind, and your knowledge of different organic molecules, what type of horse feed would you recommend to someone training Arabian horses and why?

(A) A diet high in fat and fiber so horses have greater flexibility in the selection of substrate available to meet energy demands.

(B) A diet high in sugar and starch, because this contains many soluble carbohydrates that are easy to digest.

(C) A diet high in fat and fiber, because this contains many energy sources that are easy to digest.

(D) A diet high in sugar and starch so horses have greater access to glucose substrate to meet energy demands.

Use the graph and experiment below for questions 5 and 6.

Many reef-building corals contain photosynthetic algae, called zooxanthellae, that live in their tissues. An experiment was performed to determine the rate of photosynthesis by zooxanthellae exposed to a range of temperatures from 28°C to 35°C under artificial light.

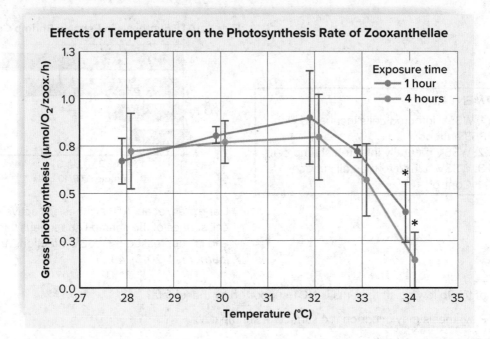

Jones, Ross J., et al. "Temperature-induced bleaching of corals begins with impairment of the CO_2 fixation mechanism in zooxanthellae." *Plant, Cell & Environment* 21.12 (1998): 1219–1230.

5. During photosynthesis, oxygen gas is released. However, recall all organisms also perform cellular respiration. In order to accurately calculate the effect of water temperature on the photosynthetic capabilities of zooxanthellae, which of the following did these scientists need to monitor in their experimental setup?

   (A) oxygen consumption

   (B) oxygen production

   (C) temperature

   (D) all of the above

6. According to the data, the zooxanthellae exposed to 34°C had significantly lower gross photosynthesis, as measured by the amount of oxygen. What must have been done in this experimental setup to confirm that this was the result of photosynthesis and not an increase in cellular respiration?

(A) Scientists did not need to measure cellular respiration rates for this experiment.

(B) Scientists measured oxygen production at night.

(C) Scientists measured oxygen production during the day and during the night.

(D) Scientists measured cellular respiration and photosynthesis using different methods.

Free Response Question

Directions: Read the questions carefully and completely. Then, plan your answer and write your response in the space provide. Write your answer out in paragraph form.

1. Proteases are enzymes that break down protein. These data show the protease activity from a newly isolated strain of bacteria.

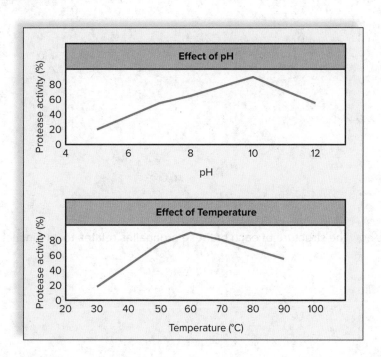

(a) **Describe** the effect of pH and temperature on the activity of the protease in these bacteria.

(b) Predict an environment where these bacteria would most likely be found. Justify your prediction.

2. The cell is the structural and functional unit of life, and contains many organelles which carry out essential cellular processes.

(a) Identify three organelles found in eukaryotic organisms and describe the critical functions of each.

(b) Explain the how the structure of each of these organelles relates to its function.

FOLLOWING *the* BIG IDEAS

AP Essential Knowledge	Chapter Section
BIG IDEA 2 **2.D.1** All biological systems from cells and organisms to populations, communities and ecosystems are affected by complex biotic and abiotic interactions involving exchange of matter and free energy.	9.4
BIG IDEA 3 **3.A.2** In eukaryotes, heritable information is passed to the next generation via processes that include the cell cycle and mitosis or meiosis plus fertilization.	9.4, 9.5
3.B.2 A variety of intercellular and intracellular signal transmissions mediate gene expression.	9.1, 9.4
3.D.3 Signal transduction pathways link signal reception with cellular response.	9.4
3.D.4 Changes in signal transduction pathways can alter cellular response.	9.4

CHAPTER OVERVIEW

Cell division is a highly regulated process. Prokaryotes and eukaryotes have different forms of cell division. Prokaryotes divide through binary fission, whereas eukaryotes undergo a more complex process in order to replicate their linear chromosomes.

9.1 The Cell Cycle

Essential Knowledge covered
3.B.2: A variety of intercellular and intracellular signal transmission mediate gene expression.

Recall It

The cell cycle is made up of two portions, interphase and the mitotic stage. During interphase, the cell grows and prepares for nuclear division. Nuclear division and the division of the cytoplasm, or cytokinesis, occurs during mitosis. In eukaryotic cells, the nucleus contains chromosomes. Each chromosome has one DNA double helix or chromatid. Chromosomes are distributed by the mitotic spindle during mitosis into two daughter cells. Agents which influence the activities of a cell are called signals, such as growth factors and cyclins. One signaling protein, p53, can stop the cycle if DNA is damaged and bring about cell death or apoptosis. Apoptosis decreases the number of somatic cells.

Review It

What is the end result of mitosis?

9.1 The Cell Cycle *continued*

Identify the major stages or checkpoints of the cell cycle in the table below.

Definition	Stage	Checkpoint
Cell growth stage, before DNA replication occurs.		
		G_1
Cells grow and DNA replicates		
	G_2	
		G_2
Mitotic stage – cell division and cytokinesis occurs.		
		M

Use It

Give an example of how cells control the cell cycle.

Syndactyly is the condition of a human born with webbed or conjoined fingers. What do you think is the cause of this defect?

9.2 The Eukaryotic Chromosome
Extending Knowledge

Recall It

A eukaryotic chromosome contains a single double helix DNA molecule but that molecule has a highly organized and complex structure. The DNA of a single chromosome contains histone molecules which are essential for the structure and compact nature of DNA. DNA is wound around a core of eight histone molecules into what is called a nucleosome. Nucleosomes are joined by linker DNA and further compacted into looped domains so it can all fit into the nucleus of a cell. Loosely coiled regions of DNA, called euchromatin, represent genes that are being transcribed. Heterochromatin are areas of the nucleus that are considered inactive.

Review It

What are histones?

Describe one advantage that chromosomes may have in condensing before cell division.

Compare and contrast euchromatin and heterochromatin.

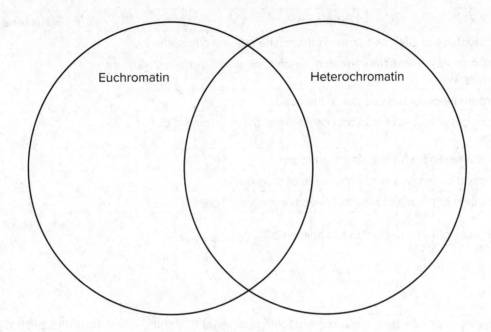

Euchromatin   Heterochromatin

9.3 Mitosis and Cytokinesis

Essential Knowledge covered
3.A.2: In eukaryotes, heritable information is passed to the next generation via processes that include the cell cycle and mitosis or meiosis plus fertilization.

Recall It

Cell division includes mitosis or nuclear division, and cytokinesis, which is division of the cytoplasm. Preparation for mitosis is called interphase. During interphase, a cell must replicate the chromosomes and duplicate most cellular organelles, including the centrosome, which will organize the spindle apparatus necessary for the movement of chromosomes. Then, during mitosis, the sister chromatids are separated and distributed to two daughter cells. Cytokinesis is the phase in which the cytoplasm divides, creating two cells. In animal cells, a cleavage furrow pinches off the daughter cells. In plant cells, a cell plate divides the daughter cells.

9.3 Mitosis and Cytokinesis *continued*

Review It

Given the definition on the left, fill in the correct structure on the right.

Definition	Structure
The tangled mass of DNA and proteins that make up a chromosome	
A newly formed plasma membrane that expands outward and fuses with an older membrane	
The microtubule-organizing center of the cell	
Protein complexes that develop on either side of the centromere during cell division	
Regions where sister chromatids are attached	
Barrel-shaped organelles in a centrosome of an animal	
An indentation of the membrane between two daughter cells in an animal	

What is an example of a haploid cell in animals?

Use It

The mosquito, *Aedes aegypti*, has a chromosome number of 6. Identify which mosquito cell is diploid and which is haploid.

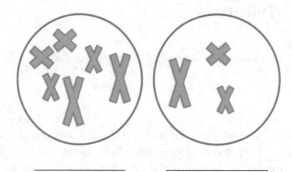

_____   _____

Draw a picture of a duplicated chromosome. Label the chromatids, the centromere, and the kinetochore.

Compare and contrast cytokinesis in animal and plant cells.

9.4 The Cell Cycle and Cancer

Essential Knowledge covered
2.D.1: All biological systems from cells and organisms to populations, communities and ecosystems are affected by complex biotic and abiotic interactions involving exchange of matter and free energy.
3.A.2: In eukaryotes, heritable information is passed to the next generation via processes that include the cell cycle and mitosis or meiosis plus fertilization.
3.B.2: A variety of intercellular and intracellular signal transmission mediate gene expression.
3.D.3: Signal transduction pathway link signal reception with cellular response.
3.D.4: Changes in signal transduction pathways can alter cellular response.

Recall It

When cells begin to divide uncontrollably, an organism has cancer. If cancer does not grow larger, it is considered benign, but if it has the ability to spread, it is malignant. Abnormal cancer cells tend to pile up on each other to form a tumor. Cancer cells may spread through the blood and lymph to start tumors elsewhere, a process called metastasis. Some tumor cells can direct the growth of new blood vessels in a process called angiogenesis. Two types of genes are often affected when cancer occurs: proto-oncogenes and tumor suppressor genes. Proto-oncogenes become oncogenes when they become mutate and cause cancer. Mutations in the ends of the chromosomes or telomeres may also cause cancer.

Review It

List five characteristics of cancer cells.

Draw a diagram showing how a cancer cell may start on the skin, become a tumor, and invade the lymphatic and blood vessel system. Be sure to label your diagram.

Use It

Determine if the following characteristic belongs to a cancer cell (CC) or a normal cell (NC).

Undergoes apoptosis

Undergoes metastasis

Has no contact inhibition

Differentiates

Has an abnormal nuclei

9.4 The Cell Cycle and Cancer *continued*

Suppose a person spends a lot of time applying a hazardous pesticide to a garden and then finds out that they have cancer. Describe a mutation to a proto-oncogene that might have been in influenced by the presence of this pesticide to cause this disease.

The person from the question above found that the cancer was actually caused by the failure of the tumor suppressor gene, p53. How is p53 related to cancer?

9.5 Prokaryotic Cell Division

Essential Knowledge covered
3.A.2: In eukaryotes, heritable information is passed to the next generation via processes that include the cell cycle and mitosis or meiosis plus fertilization.

Recall It

Binary fission is the process in which prokaryotic cells and unicellular eukaryotes split in half after replicating DNA. This allows these organisms to reproduce asexually. The structure of the prokaryotic chromosome differs from chromosomes found in eukaryotes. The prokaryotic chromosome has few associated proteins and a single loop of DNA. When binary fission occurs, the chromosome attaches to the inside of the plasma membrane and replicates. As the cell elongates, the duplicated chromosomes are pulled apart, and the cell divides, resulting in identical cells.

Review It

Fill in the chart below:

Process or Structure	Definition
Binary Fission	
	Cellular division in which offspring are genetically identical to parent
	A circular loop of DNA and proteins in a prokaryote that is not enclosed by a membrane

Use It

Determine whether or not the following statements are true or false (T/F) about the prokaryotic cell division:

Prokaryotic offspring are genetically identical to the parent

Prokaryotes lack a nucleus and other membranous organelles

Prokaryotes do not have a chromosome

After DNA replication occurs in prokaryotes, there are two genetically distinct chromosomes present

Using the diagram below, describe the steps of prokaryotic division.

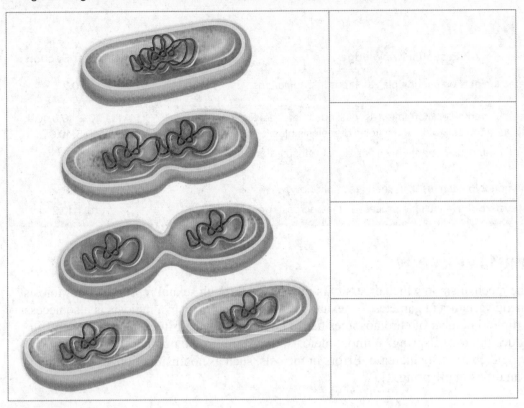

AP CHAPTER SUMMARY

Summarize It

What are stem cells and why are they important?

Compare and contrast eukaryotic and prokaryotic cell division.

How do internal and external signals regulate the cell cycle? What is the relationship between cancer and this regulation?

Following *the* Big Ideas

AP Essential Knowledge	Chapter Section
BIG IDEA 1 **1.A.2** Natural selection acts on phenotypic variations in populations.	10.2
BIG IDEA 3 **3.A.2** In eukaryotes, heritable information is passed to the next generation via processes that include the cell cycle and mitosis or meiosis plus fertilization.	10.1, 10.2, 10.3, 10.4, 10.5, 10.6
3.B.2 A variety of intercellular and intracellular signal transmissions mediate gene expression.	10.6
3.C.1 Changes in genotype can result in changes in phenotype.	10.6
3.C.2 Biological systems have multiple processes that increase genetic variation.	10.1, 10.2

CHAPTER OVERVIEW

Meiosis is the mechanism in which diversity is introduced through sexual reproduction. Meiosis provides genetic variation in gametes. In sexually reproducing organisms, meiosis is also necessary because the diploid number of chromosomes has to be reduced by half in each of the parents in order to produce diploid offspring. Without meiosis, the chromosome number of the next generation would continually increase. Errors in meiosis, such as nondisjunction, can lead to severe abnormalities in offspring.

10.1 Overview of Meiosis

Essential Knowledge covered
3.A.2: In eukaryotes, heritable information is passed to the next generation via processes that include the cell cycle and mitosis or meiosis plus fertilization.
3.C.2: Biological systems have multiple process that increase genetic variation.

Recall It

The central purpose of meiosis is to reduce chromosome numbers from diploid (2n) to haploid (n) in gametes. Gametes are the reproductive cells of sexually reproducing organisms. In humans, the gametes are sperm in males and eggs in females. Diploid cells (body cells) contain pairs of chromosomes, known as homologous chromosomes. Homologous chromosomes contain different versions of genes called alleles. In animals, after meiosis, haploid gametes (sperm or egg cells) fuse during fertilization to form new diploid cells.

Review It

What is the end result of meiosis?

10.1 Overview of Meiosis *continued*

Determine if the statements below refer to zygotes or gametes.

Statement	Zygote	Gamete
Has the haploid number of chromosomes		
Has a diploid number of chromosomes		
Undergoes development to become a mature organism		
Undergoes sexual reproduction to become a diploid cell		

Use It

Identify each chromosome, the homologous pair, the sister chromatids, and the nonsister chromatids.

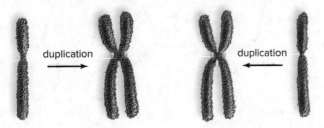

What are alleles? And where can they be found?

10.1 Overview of Meiosis *continued*

In the flow diagram below, draw the chromosomes of a cell (chromosome number = 6) undergoing DNA replication and producing daughter cells. Label when the cell is in meiosis I, and in meiosis II.

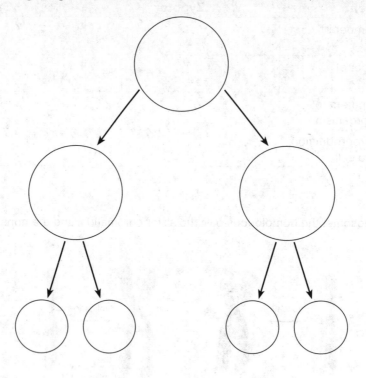

If an organism has a chromosome number of 6, how many daughter chromosomes will be present in the haploid cells?

10.2 Genetic Variation

Essential Knowledge covered
1.A.2: Natural selection acts on phenotypic variations In populations.
3.A.2: In eukaryotes, heritable information is passed to the next generation via processes that include the cell cycle and mitosis or meiosis plus fertilization.
3.C.2: Biological systems have multiple process that increase genetic variation.

Recall It

Meiosis ensures that genetic variation occurs with each generation through the processes of crossing-over and independent assortment. Crossing-over occurs during meiosis I between non-sister chromatids. The result of crossing-over is genetic recombination, which allows the chromatids that separate at the end of meiosis to have a different combination of genes. Genetic recombination is the source of variation in populations. The union of male and female gametes, or fertilization, enhances genetic variation.

10.2 Genetic Variation *continued*

Review It

Draw all possible alignments for these two homologous chromosomes during independent assortment.

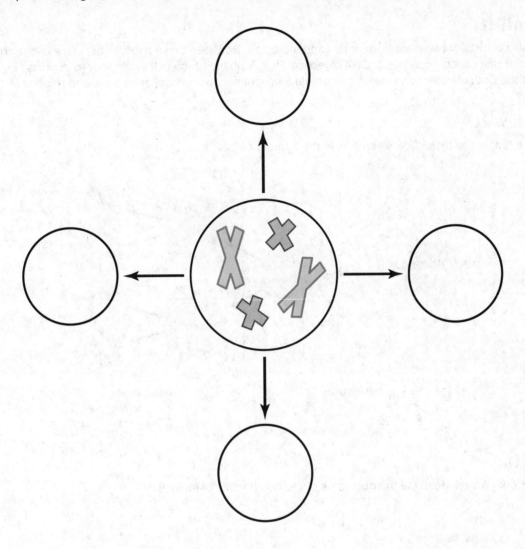

Use It

Compare and contrast crossing-over and independent assortment.

10.3 The Phases of Meiosis

Essential Knowledge covered
3.A.2: In eukaryotes, heritable information is passed to the next generation via processes that include the cell cycle and mitosis or meiosis plus fertilization.

Recall It

Meiosis consists of two phases, meiosis I and meiosis II. Between these two stages, there is a short period of rest called interkinesis. DNA is replicated in S phase of the cell cycle prior to meiosis I but not meiosis II. Both meiosis I and meiosis II contain a prophase, metaphase, anaphase, and telophase.

Review It

Place a '2n' in the diploid cells and a 'n' in the haploid cells.

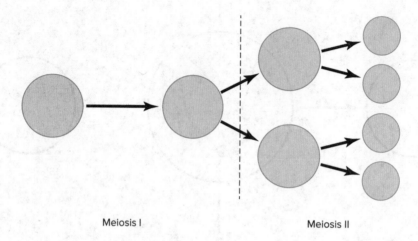

Meiosis I   Meiosis II

Use It

In your own words, describe the outcomes of the two phases of meiosis.

10.4 Meiosis Compared to Mitosis

Essential Knowledge covered
3.A.2: In eukaryotes, heritable information is passed to the next generation via processes that include the cell cycle and mitosis or meiosis plus fertilization.

Recall It

There are many similarities between meiosis and mitosis, but the outcomes of the two processes are vastly different. Meiosis requires two nuclear divisions, but mitosis requires only one nuclear division. While DNA replicates and divides during both meiosis and mitosis, in meiosis, DNA replicates once but divides twice. Meiosis produces four daughter cells, but mitosis only produces two. Meiosis produces haploid cells, and mitosis produces diploid cells. Finally, and, most importantly, meiosis produces genetically distinct daughter cells, and mitosis produces genetically identical cells to the parent cell.

10.4 Meiosis Compared to Mitosis *continued*

Review It

Determine if the following statement applies to mitosis or meiosis.

Statement	Process
Daughter cells have half the chromosome number as the parent	
Daughter cells are genetically identical to each other and the parent	
After cytokinesis, there are two daughter cells	
Requires two nuclear divisions	
No pairing of the chromosomes occur	

Place an X under the phase where the following statement applies.

Statement	Meiosis I	Meiosis II	Mitosis
In the Prophase stage, there is a pairing of chromosomes			
In Anaphase, sister chromatids separate to become daughter cells			
Four haploid cells that are not genetically identical are formed			
Two diploid daughter cells that are identical to the parent are formed			

Use It

Identify two similarities between mitosis and meiosis.

Explain why mitosis produces daughter cells with the same chromosome number but meiosis results in a reduction.

10.5 The Cycle of Life

Essential Knowledge covered
3.A.2: In eukaryotes, heritable information is passed to the next generation via processes that include the cell cycle and mitosis or meiosis plus fertilization.

10.5 The Cycle of Life *continued*

Recall It

All reproductive events that occur from one generation to the next is called an organism's life cycle. In humans, the only haploid phase of the life cycle occurs in the gametes, whereas in plants, the haploid and diploid phases are broken into generations known as gametophytes and sporophytes. In humans, the production of gametes or gametogenesis is when meiosis occurs. Meiosis in males is known as spermatogenesis and oogenesis in females. Oogenesis occurs in the ovaries, more specifically- in cells known as oocytes. A primary oocyte undergoes meiosis I to produce a second oocyte and polar body.

Review It

What is the fate of the polar body?

Insert the correct number in the statements about human spermatogenesis and oogenesis:

Spermatogenesis produces _____ viable sperm.

Oogenesis produces _____ egg and at least _____ polar bodies.

Each gamete has _____ chromosomes.

A zygote has _____ chromosomes.

Use It

Determine whether meiosis or mitosis occurs in the following human structure.

The Life Cycle of Humans		
Components	**Meiosis or Mitosis?**	**Specific Process**
Sperm		
Egg		
Zygote		—

10.6 Changes in Chromosome Number and Structure

Essential Knowledge covered
3.A.3: The chromosomal basis of inheritance provides an understanding of the pattern of passage (transmission of genes from parents to offspring).
3.B.2: A variety of intercellular and intracellular signal transmission mediate gene expression.
3.C.1: Changes in genotype can result in changes in phenotype.

Recall It

Extra or missing copies of chromosomes may result in gametes if nondisjunction occurs during meiosis I or meiosis II. If these abnormal gametes are then used in fertilization, zygotes can carry genetic abnormalities ranging from early death to mild disabilities. Deletions, duplications, inversions, and translocation are all ways in which chromosomes may be changed. Many distinct physical and physiological characteristics have been linked in humans to changes in chromosomal structure and number.

Review It

Name two aneuploid states and explain why the state occurs.

Identify the syndrome characterized by the sex chromosomes in the chart below.

Sex Chromosomes	Syndrome
XXY	
XO	
XXX	
XYY	

Use It

Describe four changes that can occur in chromosome structure which may lead to developmental abnormalities.

AP CHAPTER SUMMARY

Summarize It

How do organisms benefit from sexual reproduction in a changing environment?

How does meiosis reduce the chromosomal number from diploid to haploid and create genetically distinct daughter cells?

Compare and contrast oogenesis and spermatogenesis.

11 Mendelian Pattern of Inheritance

FOLLOWING *the* BIG IDEAS

	AP Essential Knowledge	Chapter Section
BIG IDEA 3	**3.A.1** DNA, and in some cases RNA, is the primary source of heritable information.	11.1
	3.A.2 In eukaryotes, heritable information is passed to the next generation via processes that include the cell cycle and mitosis or meiosis plus fertilization.	11.4
	3.A.3 The chromosomal basis of inheritance provides an understanding of the pattern of passage (transmission) of genes from parent to offspring.	11.2, 11.3, 11.4
BIG IDEA 4	**4.C.1** Variation in molecular units provides cells with a wider range of functions.	11.3

CHAPTER OVERVIEW

Our understanding of how genes are passed from one generation to the next started with the experiments performed by Gregor Mendel. Mendel described patterns of inheritance in pea plants; he discovered that some genes are recessive and some are dominant. He found that different traits can segregate independently of one another. Probability can often be used to determine the outcome of inheritance, but the inheritance of some traits are harder to predict for various reasons, including incomplete dominance, polygenic inheritance, pleiotropy, and environmental effect on phenotype.

11.1 Gregor Mendel

Essential Knowledge covered
3.A.3: The chromosomal basis of inheritance provides an understanding of the pattern of passage (transmission of genes from parents to offspring).

Recall It

Gregor Mendel was an Austrian monk that came up with a model in the 1860s to explain why some traits are passed on from generation to generation, and why variations exist. He called his model the *particulate theory of inheritance*. He took very careful notes and performed precise experiments with pea plants. He proposed the law of segregation and the law of independent assortment.

Review It

Name two laws of genetics that Mendel proposed.

Use It

In what way did Mendel's particulate theory disprove the blending concept?

11.2 Mendel's Laws

Essential Knowledge covered
3.A.3: The chromosomal basis of inheritance provides an understanding of the pattern of passage (transmission of genes from parents to offspring).

Recall It

Using true-breeding plants, Mendel performed a series of breeding experiments which led him to the laws of segregation and independent assortment. He performed both monohybrid and dihybrid crosses. A monohybrid cross is a cross of an organism with a single trait with an organism that is hybrid, while a dihybrid cross is that of organisms that are hybrid in two ways. Mendel coined the terms recessive to describe the trait that had disappeared in the F_1 generation, and dominant to the trait that had not. His monohybrid crosses led him to the law of segregation, which we now interpret as: an individual has two alleles for each trait, and these alleles segregate with equal probability into gametes. His dihybrid crosses led him to the law of independent assortment, which states that the members of one pair of alleles separate independently of those of another.

Review It

Identify if the pea plant is homozygous or heterozygous for being tall (T).

Genotype	Homozygous or Heterozygous
TT	
Tt	
tt	

Two parents both carry a recessive and dominant allele for the recessive genetic disorder cystic fibrosis (Cc). If they had a child with cystic fibrosis, what genotype would this child have? What is the probability they will have a child with cystic fibrosis? Use a Punnett square to illustrate your answer.

Use It

In a particular strain of roses, plants with pink flowers (P) are dominant over those with white flowers (p). If you wanted to determine the genotype of a rose plant that produced pink flowers, what test would you run?

If the results of the test gave you 50% white roses and 50% roses, what was the genotype of the original rose?

Suppose the white rose plants also carry a recessive gene for having no thorns (t). If a white thorn-less rose plant (wwtt) was crossed with a heterozygous pink rose plant (WwTt), what would the percentage of pink thorn-less roses be? Use a Punnett square to help you determine your answer.

What is the phenotypic ratio of this cross?

11.3 Mendelian Patterns of Inheritance and Human Disease

Essential Knowledge covered
3.A.3: The chromosomal basis of inheritance provides an understanding of the pattern of passage (transmission of genes from parents to offspring).
4.C.1: Variation in molecular units provides cells with a wide range of functions.

Recall It

An autosome is any chromosome other than a sex chromosome. An autosomal recessive disorder is carried on two recessive alleles (a), whereas autosomal dominant disorders are carried on the dominant allele (A). Only (aa) genotypes are affected for autosomal recessive. Autosomal dominant are affected with AA or Aa. For autosomal recessive disorders, most affected children will often have unaffected parents, and for autosomal dominant, most affected children have an affected parent. Both sexes are affected with equal frequency for either disorder. A carrier is a person who appears normal but is able to have a child with a genetic disorder.

Review It

How might a biologist show the pattern of inheritance of a particular trait within a family?

Review if the disorder you learned about in this section is either autosomal recessive (R) or autosomal dominant (D).

Disorder	Recessive or Dominant
Cystic fibrosis	
Phenylketonuria	
Osteogenesis inperfecta	
Huntington disease	
Methemoglobinemia	
Hereditary spherocytosus	

11.3 Mendelian Patterns of Inheritance and Human Disease *continued*

Use It

Examine this pedigree below. Do the affected individuals carry an autosomal dominant or recessive disorder?

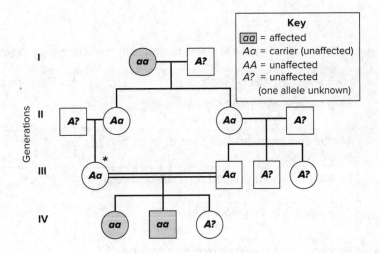

Key
aa = affected
Aa = carrier (unaffected)
AA = unaffected
$A?$ = unaffected
(one allele unknown)

What does the double line in generation III denote? How does this influence the probability of inheriting a harmful allele?

If this disorder charted in the pedigree above was methemoglobinemia, what could you tell about the blood of the affected individuals?

How is it possible that someone that might appear normal may still produce a child with the trait?

11.4 Beyond Mendelian Inheritance

Essential Knowledge covered
3.A.2: In eukaryotes, heritable information is passed to the next generation via processes that include the cell cycle and mitosis or meiosis plus fertilization.
3.A.3: The chromosomal basis of inheritance provides an understanding of the pattern of passage (transmission of genes from parents to offspring).
3.A.4: The inheritance pattern of many traits cannot be explained by simple Mendelian genetics.

11.4 Beyond Mendelian Inheritance *continued*

Recall It

Often traits are not a product of simple genetic patterns. Many traits are controlled by multiple alleles, exhibit incomplete dominance, or are a product of pleiotropic effects. Some genes are affected by the environment, and phenotypic variations are seen under different conditions. Some traits are linked to particular chromosomes, such as the X sex chromosome, even if the gene has nothing to do with gender. Males cannot be heterozygous for X-linked traits (as they only have one X chromosome).

Review It

Identify which complex pattern of inheritance explains the traits in the organism below:

Organism	Trait	Pattern of Inheritance
Fruit fly	Only males have white colored eyes	
Human	Marfan syndrome effecting lungs, eyes, and blood vessels	
Human	Has both A and B antigen on red blood cells	
Human	Skin color in the middle range	
Four-o-clock plants	Red and white flowered parents produce 1:2:1 red:pink:white flowered offspring	
Human	Polydactyly inherited in an autosomal dominant manner from a parent with only 10 digits	
Himalayan rabbit	Has dark fur only at the extremities	

Use It

A man is considered hemizogous for his red-green color blindness. What does this mean?

A person is suffering from sickle-cell anemia. Describe the disadvantages and advantages of the mutation they carry.

Summarize It

What is the relationship between genes and their passage from parent to offspring to natural selection and evolution?

How does the behavior of chromosomes during meiosis explain Mendel's laws of segregation and independent assortment?

Primroses have white flowers when grown above 32°C but red flowers when grown at 24°C. Would a simple Mendelian cross be able to predict this trait? Why or why not?

FOLLOWING *the* BIG IDEAS

AP Essential Knowledge	Chapter Section
BIG IDEA 1 **1.B.1** Organisms share many conserved core processes and features that evolved and are widely distributed among organisms today.	12.1, 12.3
BIG IDEA 3 **3.A.1** DNA, and in some cases RNA, is the primary source of heritable information.	12.1, 12.2, 12.3, 12.4, 12.5
3.B.1 Gene regulation results in differential gene expression, leading to cell specialization.	12.4
3.C.1 Changes in genotype can result in changes in phenotype.	12.2

CHAPTER OVERVIEW

DNA is the source of genetic material. DNA is a double stranded helix that contains a sugar-phosphate backbone, and complimentary nitrogenous bases. DNA replicates in a semiconservative manner and requires many different enzymes in order to synthesize new strands successfully in both prokaryotes and eukaryotes. The central dogma of molecular biology says that the flow of genetic information is from DNA → RNA → protein. DNA is transcribed into RNA, and RNA translated into protein.

12.1 The Genetic Material

Essential Knowledge covered
1.B.1:   Organisms share many conserved core processes and features that evolved and are widely distributed among organisms today.
3.A.1:   DNA, and in some cases RNA, is the primary source of heritable information.

Recall It

Several scientists in the mid twentieth century ultimately determined that DNA is the source of genetic material. Fedrick Griffith showed there was a substance that could cause heredity change through transforming nonvirulent *S. pneumoniae* bacteria into a virulent strain. Scientists Avery, MacLeod, and McCarty identified the substance responsible for the transformation in Griffith's experiment as DNA, through showing the transforming substance could be inactivated by DNA-digesting enzymes, but not by protein-digesting enzymes. Hershey and Chase later confirmed these results by using radioactive labels to mark DNA and protein in bacteriophages. The structure of DNA was then determined by Erwin Chargaff, and further elucidated by Rosalind Franklin, and Watson and Crick.

Review It

Identify the major components of DNA.

Definition	Component(s)
The two purines bases	
The two pyrimdine bases	
The shape of DNA	
The molecules that make up the "rungs" of the double helix	

List Chargaff's two rules.

Use It

Describe the seminal experiments discussed in this section and how they contributed to our understanding of the genetic material.

Describe the structure of DNA.

Explain how it is possible that, although only one of four bases can be found at each nucleotide position in DNA, there is so much variability across species.

12.2 Replication of DNA

Essential Knowledge covered
3.A.1: DNA, and in some cases RNA, is the primary source of heritable information.
3.C.1: Changes in genotype can result in changes in phenotype.

Recall It

The process of DNA replication is called semiconservative replication because the daughter strand contains one old strand and one new strand. The parental double helix serves as a template for the new strand. DNA replication begins with DNA being unwound by the enzyme DNA helicase. Single-stranded binding proteins attach to the open strands to prevent DNA from recoiling so that DNA primase can place a short primer on the strand that needs to be replicated. In eukaryotes, wherever DNA replication begins, is called a replication fork. DNA ligase binds the fragments of the replicated lagging strands back together.

Review It

Fill out the missing the enzyme or function involved in DNA replication on the chart below.

Enzyme	Function
	Places primers, begins DNA synthesis, and proof reads strands
DNA helicase	
	Mends Okazaki fragments together

12.2 Replication of DNA *continued*

What does it mean that DNA synthesized in the 5' to 3' direction?

Use It

Draw a diagram showing the enzymes involved in eukaryotic DNA replication in action. Be sure to include the following structures in your answer.

Structures
    DNA polymerase
    5' strand
    3' strand
    Okazaki fragments
    DNA helicase
    SSBs
    DNA primase
    A primer
    DNA ligase

How does DNA polymerase aid in accurately copying DNA? What would happen if it couldn't?

12.3 The Genetic Code of Life

Essential Knowledge covered
1.B.1: Organisms share many conserved core processes and features that evolved and are widely distributed among organisms today.
3.A.1: DNA, and in some cases RNA, is the primary source of heritable information.

Recall It

The central dogma of molecular biology says that the flow of genetic information is from DNA → RNA → protein. All genetic information is encoded in DNA but it relies on RNA to carry and convert this message into proteins. Messenger RNA (mRNA), transfer RNA (tRNA), and ribosomal RNA (rRNA) are specialized RNAs that make proteins through transcription and translation. During transcription, mRNA is formed, and during translation, tRNA directs the correct sequence of amino acids into proteins. As you learned in Chapter 2, proteins are constructed from subunits called amino acids. There are 20 different amino acids, which can be joined together in short or long strands. Amino acids are coded for by a series of three nucleotides called a codon. These codons are known as the genetic code, and are nearly universal to al living organisms.

12.3 The Genetic Code of Life *continued*

Review It

In this section, you were introduced to three different types of RNA: mRNA, tRNA, and rRNA. Describe their similarities and differences in the chart below.

	Similar Traits	Unique Function
mRNA		
tRNA		
rRNA		

List the three stop codons and the one start codon.

What is the central dogma?

Use It

Illustrate how DNA flows from the nucleus to become a polypeptide.
Be sure to include the words *transcription*, *translation*, and *mRNA* in your answer.

How does the universal nature of the genetic code link all living organisms?

12.4 First Step: Transcription

Essential Knowledge covered
3.A.1: DNA, and in some cases RNA, is the primary source of heritable information.
3.B.1: Gene regulation results in differential gene expression, leading to cell specialization.

Recall It

Transcription is the process in which DNA is transcribed to mRNA. Transcription begins when RNA polymerase attaches to the promoter of a gene until it reaches a stop sequence. The mRNA transcript is then processed through removal of introns, splicing of exons, and an addition of a poly-A tail and a 5' cap. Then, the mRNA is ready to leave the nucleus for translation.

12.4 First Step: Transcription *continued*

Review It

Describe three reasons why the promoter is important to DNA transcription.

List two key functions of introns.

Identify the three major modifications of mRNA that occur before it is ready to leave the nucleus for protein translation.

Use It

Process this pre-mRNA so it can go deliver a message!

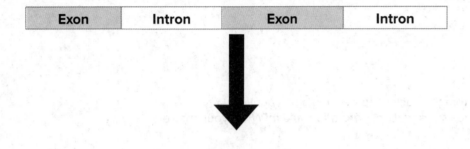

What is alternative mRNA splicing, and how does it aid in phenotypic flexibility?

12.5 Second Step: Translation

Essential Knowledge covered
3.A.1: DNA, and in some cases RNA, is the primary source of heritable information.

Recall It

The second major step in protein synthesis is translation. During translation, a processed mRNA undergoes a series of steps known as initiation, elongation, and termination to produce a protein. Initiation brings all the necessary translation components together: ribosomal subunits, mRNA, and initiator tRNA. The tRNA pairs the mRNA with the start codon (AUG) in prokaryotes. Next, the polypeptide elongates, one amino acid at a time. A stop codon denotes the termination of polypeptide synthesis.

12.5 Second Step: Translation *continued*

Review It

Identify the three major phases of translation.

Definition	Phase
The step that brings all the translation components together	
Polypeptides increase in length, one amino acid at a time	
Finished polypeptide and assembling components separate	

Determine if the following statements refer to a transfer RNA (tRNA) or ribosomal RNA (rRNA).

Transfers amino acids to the ribosomes

Is packaged in two subunits of unequal size

Becomes "charged" when an amino acid is attached

Contains an anticodon

Aids in the creation of a peptide bond between amino acids

Use It

How does a tRNA molecule correctly align amino acids in the order predetermined by DNA?

What are the three binding sites on the small subunit of the ribosome, and what role do they play in polypeptide synthesis?

How does termination occur?

12.5 Second Step: Translation *continued*

Illustrate what happens to mRNA once it moves into the cytoplasm and how it becomes a protein. Be sure to include tRNA, ribosomes, and anticodons in your drawing.

AP CHAPTER SUMMARY

Summarize It

Compare and contrast prokaryotic and eukaryotic DNA replication.

Describe the two steps of how DNA is converted into proteins.

13 Regulation of Gene Expression

FOLLOWING *the* BIG IDEAS

AP Essential Knowledge	Chapter Section
BIG IDEA 1 **1.C.3** Populations of organisms continue to evolve.	13.3
BIG IDEA 3 **3.B.1** Gene regulation results in differential gene expression, leading to cell specialization.	13.1, 13.2, 13.3
3.B.2 A variety of intercellular and intracellular signal transmissions mediate gene expression.	13.1, 13.2
3.C.1 Changes in genotype can result in changes in phenotype.	13.2, 13.3
BIG IDEA 4 **4.A.2** The structure and function of subcellular components, and their interactions, provide essential cellular processes.	13.1, 13.2

CHAPTER OVERVIEW

There are many mechanisms located within cells which control the expression of genes. In prokaryotes, gene regulation is most often controlled through transcription. Pathways called operons may be regulated by positive or negative feedback systems. In eukaryotes, gene expression is more complex; regulatory genes, regulatory elements, and transcription factors can all operate simultaneously. Gene regulation in eukaryotes also occurs at different levels, such as in chromatin structure, translational control, and post translational modifications. Changes in DNA through mutations can lead to various affects in gene expression.

13.1 Prokaryotic Regulation

Essential Knowledge covered
3.B.1: *Gene regulation results in differential gene expression, leading to cell specialization.*
3.B.2: *A variety of intercellular and intracellular signal transmission mediate gene expression.*
4.A.2: *The structure and function of subcellular components, and their interactions, provide essential cellular processes.*

Recall It

In prokaryotes, control of gene expression is often regulated through operons. Operons can be positive controlled or negative controlled. Positive control increases the frequency of transcription, and it is mediated by regulatory proteins called activators. Negative control is mediated by proteins called repressors, which interfere with transcription. Repression occurs when gene expression is turned off in the presence of a substrate, such as in the trp operon. The *trp* operon is repressed when tryptophan, acting as a corepressor, binds to the repressor, altering its conformation such that it can bind to DNA and turn off the operon. Induction occurs when gene expression is turned on in the presence of a substrate, such as in the lac operon. When lactose is present, allolactose binds to the lac operon repressor, and the operon is turned on.

13.1 Prokaryotic Regulation *continued*

Review It

Identify the major components of prokaryotic gene regulation.

Definition	Component
An enzyme that brings out the expression of a gene	
A short portion of DNA located before the structural genes	
An area outside the operon which encodes for a repressor protein	
Genes that code for enzymes and proteins involved in a the metabolic pathway of an operon	
A molecule that binds to a repressor and prevents a gene from being expressed	
A short sequence of DNA where RNA polymerase first attaches to begin the transcription of an operon	

List the four components of an operon

Use It

The following diagram illustrates the *trp* operon which regulates gene expression in *E. coli* with the amino acid tryptophan. Study the diagram below then answer the following questions.

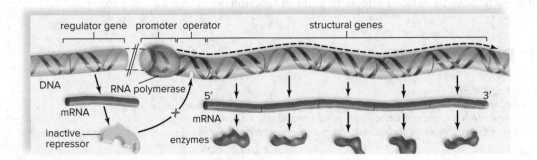

What are the products that this operon regulates?

Is this operon in the "on" or "off" position (i.e. allowing for gene expression or shutting off the expression of gene expression) without tryptophan present.

Where does tryptophan bind?

What happens to the operator and promoter in the operon when tryptophan binds?

13.1 Prokaryotic Regulation *continued*

Why is the *trp* operon considered a repressible operon?

13.2 Eukaryotic Regulation

Essential Knowledge covered
3.B.1: Gene regulation results in differential gene expression, leading to cell specialization.
3.B.2: A variety of intercellular and intracellular signal transmission mediate gene expression.
3.C.1: Changes in genotype can result in changes in phenotype.
4.A.2: The structure and function of subcellular components, and their interactions, provide essential cellular processes.

Recall It

There are several layers of gene regulation in eukaryotes. Gene regulation can occur through chromatin structure, chemical modifications of chromosomes (DNA methylation), transcription factors, and even small RNA molecules. Small-interfering RNAS and microRNAs can interfere at the transcriptional level. Posttranscriptional control can occur through variations in mRNA splicing. Translational control can also affect mRNA translation. This complex variation in gene expression leads to differential gene expression within different cells of the body.

Review It

Describe five mechanisms that can control gene expression in eukaryotes.

Based on the description of genetic control, determine the name of the control and where it takes place.

Description	Name of control	Location it takes place
A missing transcription factor		
Incorrect folding		
A missing 5' cap		
Pre-mRNA is spliced alternatively		

Use It

How does a protein enter a proteasome and what happens once it enters?

How might scientists be able to use sRNA as therapeutic agents to stop certain diseases?

13.3 Gene Mutations

Essential Knowledge covered
1.C.3: Populations of organisms continue to evolve.
3.B.1: Gene regulation results in differential gene expression, leading to cell specialization.
3.C.1: Changes in genotype can result in changes in phenotype.

Recall It

Mutations are heritable changes in genetic material. Mutations can be either induced or spontaneous, both of which are random. There are many forms of mutations, and they may have a positive, negative, or no effect on the phenotype of an individual. Point mutations affect a single site in the DNA. Point mutations can result in the addition, deletion, or substitution of one base for another. Base substitutions, silent mutations, nonsense mutations, frameshift mutations, or triplet repeat expansion mutations can arise from changes in base pairs. Many human disorders are the result of both DNA mutations.

Review It

Define gene mutation.

Place an X under the mutation where the following statement may apply.

	Induced	Spontaneous	Point	Frameshift
Happens for no apparent reason				
Occurs when a nucleotide is added or deleted from DNA				
Results from exposure to chemicals or radiation				
Is a permeant change in the sequence of DNA				
Involve a change of a single nucleotide				

Use It

Compare and contrast how mutations in a tumor suppressing gene and a proto- oncogene can lead to cancer.

Summarize It

Can genetic mutations ever be beneficial? Why or why not?

Describe the similarities and difference in gene regulation between prokaryotes and eukaryotes.

Following *the* Big Ideas

	AP Essential Knowledge	Chapter Section
BIG IDEA 1	**1.C.3** Populations of organisms continue to evolve.	14.2
BIG IDEA 3	**3.A.1** DNA, and in some cases RNA, is the primary source of heritable information.	14.1, 14.2, 14.3, 14.4

Chapter Overview

Biotechnology is the manipulation of living organisms to create a desired process or product. Genetic engineering has allowed humans to manipulate the heritable information of living organisms. New genes can be inserted into different organisms; creating ways to fight diseases, and even creating new processes or behaviors in organisms. Genomics has allowed humans to sequence the entire human genome, providing a better understanding of the genes and noncoding regions of our chromosomes. Comparative genomics allows us to look for differences and similarities in the DNA of many different species and our evolutionary relationships.

14.1 DNA Cloning

Essential Knowledge covered
3.A.1: DNA, and in some cases RNA, is the primary source of heritable information.

Recall It

Cloning is a process in which genetically identical copies of DNA, cells, or organisms are produced. Using recombinant DNA and plasmids as a vector, DNA from different sources can be combined. Restriction enzymes and DNA ligase are required to cut and bind the new piece of DNA into the plasmid. The polymerase chain reaction (PCR) is a powerful tool that was developed to amplify a specific sequence of DNA in a very short period of time. Amplified DNA can be used in DNA fingerprinting.

Review It

Identify the technology below based on the description provided.

Description	Technology
Can identify and distinguish between individuals bases on variation in their DNA	
Using cloned genes to modify a person	
Amplifies a small piece of DNA by making millions of copies using DNA polymerase	
Producing genetically identical copies of DNA, cells, or organisms through asexual means	
Contains DNA from two or more different sources and is carried in a vector	
Separates DNA fragments according to size	

14.1 DNA Cloning *continued*

Use It

Someone stole the cookies from the cookie jar, but they made a critical mistake: they left their DNA behind, all over an empty glass of milk at the scene of the crime. Taking cookie theft very seriously, the owner had a DNA profile run and tested against the DNA on the three suspects: Who Me, Yes You, and Couldn't Be. With this information in mind, answer the following question

There was only a trace amount of DNA recovered from the empty glass. How did scientists increase the amount so they could perform more accurate tests on it? Describe the process in detail.

The scientist working on the case only amplified a portion of the DNA which codes for the human CookieMnstr gene. What tool did the scientist use to do this?

Draw a line under all the biotechnology tools used in the next sentence:The CookieMnstr gene was broken into short tandem repeat sequences using a restriction enzyme and separated using gel electrophoresis for analysis.

Looking at the results, who did it and how can you tell?

	Glass DNA	Who Me	Yes You	Couldn't Be
Base repeat units				
DNA band pattern				

14.2 Biotechnology Products

Essential Knowledge covered
1.C.3: Populations of organisms continue to evolve.
3.A.1: DNA, and in some cases RNA, is the primary source of heritable information.

14.2 Biotechnology Products *continued*

Recall It

Biotechnology has been used to create many different types of products. Biotechnology products are products produced by transgenic organisms. Transgenic bacteria, plants, or animals which now contain foreign DNA are known as genetically modified organisms. Some of these organisms have been engineered to create pharmaceuticals. This is known as "gene pharming."

Review It

Determine if the following statements about biotechnology products are true (T) or false (F).

Bioengineering can enhance the natural abilities of bacteria to breakdown toxins for their use in bioremediation.

Humans contain plasmid vectors which are directly inserted into bacteria for the production of human pharmaceuticals.

Only animals can be cloned.

Golden Rice was engineered to contain human genes.

Animal cloning is a difficult process with a low success rate.

Use It

Describe how scientists used mice to discover the function of the human section of DNA called *SRY*.

14.3 Gene Therapy

Essential Knowledge covered
3.A.1: DNA, and in some cases RNA, is the primary source of heritable information.

Recall It

Gene therapy is the manipulation of an organism's genes that has been extended to humans to treat various diseases. Gene therapy can be either ex vivo or en vivo; a process which either takes place outside the body or inside the body. Examples of ex vivo therapy include removal and treatment of cells, such as bone marrow or liver cells, before being returned to the patient. In vivo gene therapy gene treatment is delivered directly to the patient through injection or aerosol spray.

Use It

Compare and contrast ex vivo and in vivo gene therapy. Be sure to define each method, provide examples of targeted conditions, and highlight the delivery mechanisms.

14.4 Genomics

Essential Knowledge covered
3.A.1: DNA, and in some cases RNA, is the primary source of heritable information.

Recall It

Genomics is the study of the genome; the complete genetic makeup of an organism. The human genome was completely sequenced by the Human Genomic Project. This project allowed scientists to learn a great deal about the number and types of genes humans have, as well as, the large amounts of noncoding DNA that our genome harbors. The vast majority of the human genome is made up of intergenic sequences and interspersed repeats, DNA found between genes that are often repetitive in nature. Transposons are specific DNA sequences which have the ability to move around and within chromosomes. Functional and comparative genomics allows us to determine the role of the genome and how genomes of different species compare, respectively. Proteomics, on the other hand, is the study of the structure, function, and interaction of only proteins in an organism. These fields would not be possible without bioinformatics; the application of computer technology to vast datasets which are created when sequencing entire genomes or proteomes of a single organism.

Review It

Identify the field you would need to solve the following question.

Question	Technology
What type of proteins are present in the human eye?	
How many genes are in the genome of a panda?	
What has more genes, an orchid or a potato?	
What genes in the brain of a mouse are turned on when it is sleeping?	

Use It

Describe how a transposon may turn the kernel of corn a different color.

14.4 Genomics *continued*

It was once thought that the Human Genome Project would discover over 100,000 genes when the project first began but this estimate was revised to 21,000–23,000 as more data became available. What accounts for all the "missing genes"?

AP CHAPTER SUMMARY

Summarize It

How can genetic engineering techniques be used to benefit human society?

How can DNA analysis and genome comparison allow us to better understand the evolution of species?

How can the manipulation of DNA by humans affect the evolution of species, and what are the ethical, medical, or social implications?

UNIT TWO: GENETIC BASIS OF LIFE

Chapter 9 The Cell Cycle and Cellular Reproduction • Chapter 10 Meiosis and Sexual Reproduction • Chapter 11 Mendelian Patterns of Inheritance • Chapter 12 Molecular Biology of the Gene • Chapter 13 Regulation of Gene Expression • Chapter 14 Biotechnology and Genomics

Multiple Choice Questions

Directions: For each question or incomplete statement chose one of the four suggested answers or completions listed below.

1. A particular plant extract used in herbal remedies was tested for its effect on cell growth. Cells were treated with the plant extract at three different concentrations. After exposure, the cells were stained and the different phases of mitotic division were counted in a total of 500 cells for each treatment.

Number of cells in different phases					
Treatment	Interphase	Prophase	Metaphase	Anaphase	Telophase
Control	384	56	11	39	10
1 M	375	61	12	43	9
10 M	381	53	9	46	11
100 M	407	71	19	1	2

The results of this study best support which of the following hypotheses?

(A) At high concentrations, this plant extract interferes with chromosomal pairing.

(B) This plant extract has no effect on cell division.

(C) At high concentrations, this plant extract interferes with spindle fiber disassembly.

(D) At high concentrations, this plant extract interferes with nuclear envelope fragmentation.

2. In a cross between a true-breeding, red-flowered four-o'clock plant strain and a true-breeding, white-flowered strain, the offspring (F_1) have pink flowers. When the pink plants self-pollinate, the offspring of F_1 have a phenotypic ratio of 1 red-flowered: 2 pink-flowered; 1white-flowered. The outcome of the initial cross can best be described by which of the following?

(A) Incomplete dominance

(B) The blending theory of inheritance

(C) X-linked Inheritance

(D) Polygenic inheritance

3. A common scientific technique is to insert a small piece of DNA into a plasmid. Which enzyme, also critical to DNA replication, would be necessary for this process?

(A) RNA polymerase

(B) topoisomerase

(C) DNA ligase

(D) helicase

4. A Fragile-X syndrome is a genetic disorder leading to moderate to severe mental and physical disabilities. Individuals with Fragile-X have been shown to carry a mutated gene, now known as FRM-1.

One particular patient showed to carry the following mutation:

Patient	AAG	CTG	AAT	CAG	GAG
Control	AAG	CTG	ATT	CAG	GAG

According to this data, this particular patient with Fragile-X carries which type of DNA mutation:

(A) A frameshift mutation

(B) A base addition

(C) A base deletion

(D) A point mutation

5. Hemophilia is often the result of this type of chromosomal mutation illustrated in the image below:

(A) A deletion

(B) A translocation

(C) An inversion

(D) A translocation

6. If the gene sequence of alleles C and D on the left chromosome in the illustration above is: 5′ TAGCCT 3′, then the sequence of these alleles on the mutated chromosome would be:

(A) 5′ TCCGAT 3′

(B) 5′ AGGCTA 3′

(C) 3′ AGGCTA 5′

(D) 5′ TAGCCT 3′

Free Response Questions

Directions: Read the questions carefully and completely. Then, plan your answer and write your response in the space provide. Write your answer out in paragraph form.

1. Mitosis and meiosis are two processes which seem very similar but have two distinctly different functions.

(a) **Construct** an explanation as to how DNA in chromosomes are transmitted to the next generation via mitosis.

(b) Construct an explanation as to how DNA in chromosomes are transmitted to the next generation via meiosis plus fertilization.

(c) Identify the differences between your two explanations.

2. The diagram below shows the pedigree pattern of the blood disorder hemophilia in the royal family of Europe.

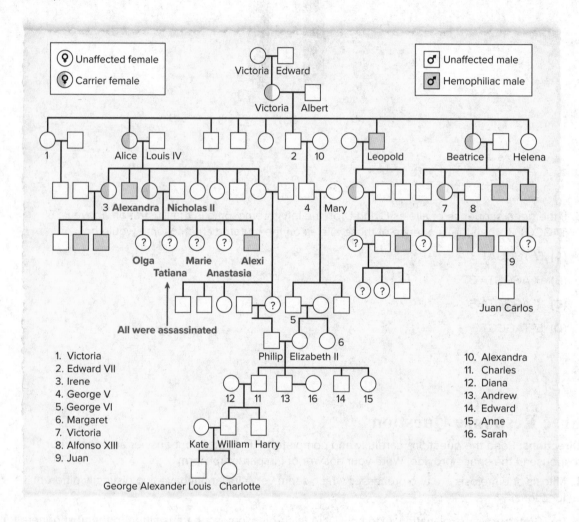

1. Victoria
2. Edward VII
3. Irene
4. George V
5. George VI
6. Margaret
7. Victoria
8. Alfonso XIII
9. Juan

10. Alexandra
11. Charles
12. Diana
13. Andrew
14. Edward
15. Anne
16. Sarah

(a) Identify what type of inheritance the hemophilia allele is, and **explain** your answer.

(b) Queen Victoria was a carrier of this allele. What were the chances of her sons having the disorder or her daughters being carriers?

(c) Why are there no hemophiliacs in the present British royal family?

15 Darwin and Evolution

FOLLOWING *the* BIG IDEAS

AP Essential Knowledge	Chapter Section
BIG IDEA 1 **1.A.1** Natural selection is a major mechanism of evolution.	15.1, 15.2, 15.3
1.A.2 Natural selection acts on phenotypic variations in populations.	15.2
1.A.4 Biological evolution is supported by scientific evidence from many disciplines, including mathematics.	15.2, 15.3
1.C.3 Populations of organisms continue to evolve.	15.2
BIG IDEA 4 **4.B.1** Interactions between molecules affect their structure and function.	15.2

CHAPTER OVERVIEW

Darwin's observation of finches on the Galapagos Islands in the 19th century provided the foundation for today's understanding of evolution. Changes in species over time by natural selection leads to genetic and phenotypic difference. Evolution is the central theory of biology, and is supported by evidence from many disciplines: paleontology, biogeography, genetics, mathematics, and biochemistry.

15.1 History of Evolutionary Thought

Essential Knowledge covered
1.A.1: Natural selection is a major mechanism of evolution.

Recall It

Before Darwin's Theory of Evolution, there were many other ideas that were proposed to explain the diversity of life we see on Earth. Early discoveries of fossils lead some to believe in catastrophism: that God had continuously created new species to repopulate the world after worldwide catastrophes. Early geologists James Hutton and Charles Lyell helped explain how geological changes occur more slowly than sudden catastrophes. Biologist Jean-Baptiste de Lamark suggested the idea of inheritance of acquired characteristics – that the environment produces physical changes in organisms that are inheritable. Charles Darwin had an alternate theory that, after reading a work by economist Thomas Malthus, would revolutionize the history of evolutionary thought.

Review It

Concepts that were fundamental in the development of evolutionary thought are listed below. Provide a definition for the following:

Concept	Definition
catastrophism	
extant	
vestigial structures	
paleontology	
evolution	
strata	
uniformitarianism	
inheritance of acquired characteristics	

Use It

Describe how paleontology and geology are important to understanding evolution.

How did Lamarck's theory influence Darwin's idea of natural selection?

How did Darwin use Thomas Malthus' work to describe changes in animal population?

15.2 Darwin's Theory of Evolution

Essential Knowledge covered

1.A.1: Natural selection is a major mechanism of evolution.
1.A.2: Natural selection acts on phenotypic variations in populations.
1.A.4: Biological evolution is supported by scientific evidence from many disciplines, including mathematics.
1.C.3: Populations of organisms continue to evolve.
4.B.1: Interactions between molecules affect their structure and function.

15.2 Darwin's Theory of Evolution *continued*

Recall It

Charles Darwin took a five-year trip to the Southern Hemisphere in 1831. There he observed massive geological changes, fossilized remains, and many unique organisms that were similar but very different to the ones he was use to in his home land in England. This trip inspired him to write his famous theory of evolution: that all life on Earth descends from a common ancestor, and natural selection is a mechanism for evolutionary change. This idea would revolutionize the field of biology and has become a corner stone of the field today. Evolutionary fitness is measured by the number of surviving offspring an individual has to produce another generation. In order for natural selection to occur there must be four characteristics met: (1) Organisms must possess heritable variation; (2) Organisms must compete for available resources; (3) Individuals differ in terms of reproductive success; and (4) Organisms become adapted to conditions as the environment changes. Human-controlled breeding of organisms for a particular trait is known as artificial selection.

Review It

Identify the concept which the example illustrates.

Concept	Example
	A toad has the ability to look like a stone and blend in with the forest floor
	Darwin studied animals in the Southern Hemisphere and in the Northern Hemisphere
	Gray mice living on gray rocks have the greatest number of offspring which survive
	A golden retriever has big floppy ears and a waggy tail

List the four observations that natural selection is based upon.

Use It

A doctor prescribes you an antibiotic for a bacterial infection and says, "It's very important that you finish the entire course, even if you think you feel better." What is the evolutionary rationale behind this advice?

The Industrial Revolution in Great Britain caused a lot of smog and soot to pollute the air. How did this affect the populations of peppered moths living in the area?

15.2 Darwin's Theory of Evolution *continued*

Describe two reasons why studying geology helped Darwin form his theory of evolution.

Darwin studied many organisms on his trip aboard the HMS Beagle, including groups of finches. The finches all had very different beaks. Use the graph below to describe one possible way in which beaks may change in a finch.

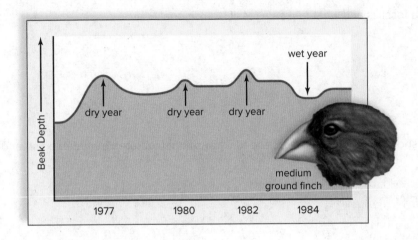

15.3 Evidence for Evolution

Essential Knowledge covered
1.A.1: Natural selection is a major mechanism of evolution.
1.A.4: Biological evolution is supported by scientific evidence from many disciplines, including mathematics.

Recall It

Scientific evidence from many disciplines including: mathematics, biogeography, geology, paleontology, chemistry, and genetics, provides support for Darwin's theory of evolution. The fossil record provides direct evidence of past life. Transitional fossils allow us to retrace the evolution of organisms over long periods of time. The history of the Earth, as seen through biogeography, shows how populations isolated to different areas of the world through geological change have adapted differently to their environments. Anatomical, chemical, and genetic similarities are evident throughout all domains of life.

Review It

Give two examples of homologous structures.

15.3 Evidence for Evolution *continued*

Give two examples of analogous structures.

Name the type of genes which orchestra the development of the body plan in all animals. Are these genes homologous or analogous?

Describe what a fossil is and what makes a fossil "transitional."

Use It

Draw a picture showing how *Tiktaalik roseae* has transitional features between a fish and a tetrapod.

Determine whether or not the following structures are homologous (H) or analogous (A).

The tentacles of a star-nosed mole and the tentacles of an octopus

The tail of an airplane and the tail of a bird

The bill of a duck and the beak of a sparrow

The bones in the fin of a whale and the bones in the leg of a cow

A gardener we met back in Section 9.4 had been using pesticides. This particular garden had pesticides applied to it for many generations to control insects in their gardens. One day, the gardener noticed that the insects on the plants no longer died when the pesticide was applied. Describe what may have happened to the insect populations in terms of evolution.

AP CHAPTER SUMMARY

Summarize It

Why are grass snakes green? Give an example of how Cuvier, Lamarck, and Darwin might answer this question.

Compare and contrast artificial and natural selection.

How do random mutations in DNA and genetic variation result in phenotypic variations that are subject to natural selection?

Describe how evolution can be observed and tested.

FOLLOWING *the* BIG IDEAS

AP Essential Knowledge	Chapter Section
1.A.1 Natural selection is a major mechanism of evolution.	16.1, 16.2, 16.3
1.A.2 Natural selection acts on phenotypic variations in populations.	16.2, 16.3
1.A.3 Evolutionary change is also drive by random processes.	16.1
1.A.4 Biological evolution is supported by scientific evidence from many disciplines, including mathematics.	16.1

BIG IDEA 1

CHAPTER OVERVIEW

Evolution within populations is measured as a change in allele frequencies over generations. Natural selection acts on trait variation, and trait variation is determined by genes. Whether or not a trait gives an advantage depends upon the environment. Genes, traits, environment, and natural selection are all involved in microevolution. There are three general types of natural selection: stabilizing selection, directional selection, and disruptive selection.

16.1 Genes, Populations, and Evolution

Essential Knowledge covered
1.A.1: Natural selection is a major mechanism of evolution.
1.A.3: Evolutionary change is also drive by random processes.
1.A.4: Biological evolution is supported by scientific evidence from many disciplines, including mathematics.

Recall It

Population genetics studies diversity of populations at the genetic level. A population is a group of individuals of the same species living together in the same geographic region. Evolutionary change within populations is referred to as microevolution, and geneticists look for these changes in the alleles of all genes of individuals in a population or in the gene pool. The percentage of each allele in a population's gene pool is known as the allele frequency. When this percentage is stable it is in Hardy-Weinberg equilibrium. For a population to be in Hardy-Weinberg equilibrium there must be no mutation, no migration, a large gene pool, random mating, and no selection. The frequency of a non-evolving population can be described by a mathematical model called the Hardy-Weinberg principle. As the movement of alleles within a population becomes more and more different over time, reproductive isolation can occur which means parts of the populations can no longer interbreed. Inbreeding does not affect the frequency of alleles alone but can have a significant impact on the genotype and phenotype of an individual. A change in allelic frequency due to chance events is due to genetic drift. There are two types of genetic drift: (1) the bottleneck effect, and (2) the founder effect. Nonrandom matting or assertive mating can skew the frequency of prevalent genotypes.

Review It

Describe the five conditions of the Hardy-Weinberg principle that need to be met in order for a population to be at equilibrium.

Compare and contrast a bottleneck effect and a founder effect.

Use It

Estimate the genotype frequencies of a population at Hardy-Weinberg equilibrium given the following allele frequencies: 0.30 H, 0.70 h

If the above population's F_2 generation has a genotypic frequency of HH=0.7, Hh=0.22, hh=0.08, can you determine if evolution occurred? Why or why not?

16.2 Natural Selection

Essential Knowledge covered
1.A.1: Natural selection is a major mechanism of evolution.
1.A.2: Natural selection acts on phenotypic variations in populations.

Recall It

Natural selection is the foundation of Darwin's theory of evolution. There are three general types of natural selection: (1) stabilizing selection, (2) directional selection, and (3) disruptive selection. In each type of natural selection, the average value of a phenotype changes over time. Sexual selection is the adaptive change in males and females that leads to increased ability to secure a mate. Sexual dimorphism occurs when males and females evolve substantially different phenotypes due to sexual selection.

16.2 Natural Selection *continued*

Review It

Identify the missing concept or definition in the chart below:

Concept	Definition
cost-benefit analysis	
	the ability to produce surviving offspring
	adaptive changes in males and females that lead to an increased ability to secure a mate
sexual dimorphism	
	controlled by many genes

Describe the three types of natural selection shown below in terms of change in phenotypes.

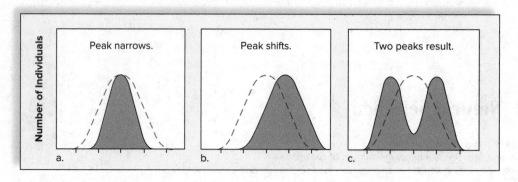

Use It

Identify the type of natural selection described in the examples below.

Example	Selection Type
Human infants with an intermediate weight have a better chance of survival	
A population of fish feed on detritus in deep water but start feeding on flies when in shallow water	
Asters growing in a tundra have small leaves and flowers	
A drab female bird mates with a brightly colored male	

16.2 Natural Selection *continued*

Explain in terms of natural selection, why British land snails tend to have two distinct phenotypes.

Describe two ways sexual selection may occur.

16.3 Maintenance of Diversity

Essential Knowledge covered
1.A.1: Natural selection is a major mechanism of evolution.
1.A.2: Natural selection acts on phenotypic variations in populations.

Recall It

Genetic diversity is maintained in populations through mutations and recombination events. Gene flow among small populations can introduce new alleles, and natural selection can also result in variation. Recessive alleles are often maintained even if the homozygous recessive condition is detrimental or selected against in diploid organisms. Stabilizing selection can act upon these alleles if they also provide another advantage in a particular environment.

Review It

When does the heterozygote advantage occur? And how does this affect the maintenance of diversity?

Describe another mechanism which promotes the maintenance of diversity.

Use It

Explain why Sickle-cell disease remains prevalent in Africa, despite the devastating affects it can have on an individual.

Summarize It

What is the connection between change in the environment and change in allele frequencies?

How can the Hardy-Weinberg mathematical model be used to analyze genetic drift and effects of selection in the evolution of populations?

17 Speciation and Macroevolution

FOLLOWING *the* BIG IDEAS

AP Essential Knowledge	Chapter Section
BIG IDEA 1 **1.C.1** Speciation and extinction have occurred throughout the Earth's history.	17.1, 17.2, 17.3
1.C.2 Speciation may occur when two populations become reproductively isolated from each other.	17.2, 17.3
1.C.3 Populations of organisms continue to evolve.	17.1, 17.3

CHAPTER OVERVIEW

Over time, populations accumulate differences large enough to become a new species. This is known as speciation. Biologists use several concepts to define species, and there are many modes of speciation. Some species evolve gradually, and other evolve more rapidly.

17.1 How New Species Evolve

Essential Knowledge covered
1.C.1: Speciation and extinction have occurred throughout the Earth's history.
1.C.2: Speciation may occur when two populations become reproductively isolated from each other.
1.C.3: Populations of organisms continue to evolve.

Recall It

Macroevolution is evolution on a large scale and involves speciation: the splitting of one species into two or more. Scientists explain species through the phylogenetic species concept and the biological species concept. The phylogenetic species concept does not rely on morphology alone to define a species—nucleotide sequences are used to recognize and aid in determining the phylogeny of species. Different species are ultimately defined if two organisms cannot mate or if they produce sterile offspring. Prezygotic and postzygotic mechanisms lead to reproductive isolation between two species.

Review It

Describe the different types of species concepts:

Concept	Definition
Morphological Species Concept	
Evolutionary Species Concept	
Phylogenetic Species Concept	
Biological Species Concept	

17.1 How New Species Evolve *continued*

List five prezygotic isolating mechanisms.

List two postzygotic mechanisms.

Use It

A horse and a donkey can mate and produce offspring. Does this mean that they are the same species?

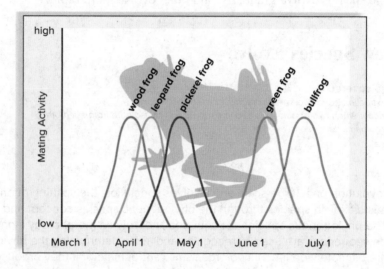

The figure above illustrates the breeding times of different species of frogs in the genus *Rana*. What type of mechanism may have led to their speciation?

Scientists found another frog living in this region that looked exactly like the leopard frog but had a very different sounding call.

Scientists would classify this species as what kind of species?

How would they determine if it was a new species?

What type of event may have led to its speciation, assuming it preferred the same habitat and reproduced at the same time?

17.2 Modes of Speciation

Essential Knowledge covered
1.C.1: Speciation and extinction have occurred throughout the Earth's history.
1.C.2: Speciation may occur when two populations become reproductively isolated from each other.

Recall It

There are many ways in which speciation occurs. Allopatric speciation occurs when populations becomes separated through a geographical or physical barrier and adapt to their new environments. Sympatric speciation occurs in the absence of a physical barrier. Divergences often occur through behavioral changes due to microhabitats or through polyploidy in plants. Adaptive radiation occurs more rapidly. During adaptive radiation, a single ancestral species gives rise to many new species often following the removal of a competitor or due to a change in the environment. Convergent evolution may occur in two different species that are affected by similar selective pressures.

Review It

Describe the different models of evolution:

Description	Model
Two populations of salmon develop different body size and shape because of geographical isolation	
Cichlid fishes living in open water need to feed on different food than cichlids living on the coastline	
Silversword plants in Hawaii have adapted to live in lava fields as well as dry and wet environments	

Describe convergent evolution, and provide an example.

Use It

Imagine a small population of furry creatures colonized a chain of islands, and their descendants spread out to occupy many different niches. How might populations become different as a result of adaptive radiation?

Explain how the extinction of the dinosaurs 66 million years ago allowed for the diversification of mammals.

17.3 Principles of Macroevolution

Essential Knowledge covered
1.C.1: Speciation and extinction have occurred throughout the Earth's history.
1.C.2: Speciation may occur when two populations become reproductively isolated from each other.
1.C.3: Populations of organisms continue to evolve.

Recall It

Evidence of both punctuated equilibrium and gradualism are seen in the fossil record. Rapid change can occur in species due to changes in regulatory genes, such as the *Hox* genes. Random mutations can lead to variation in traits that allow organisms to adapt to particular environments. All adaptations are subject to natural selection. Changes, whether slow or fast, will continue to occur, causing speciation to continue.

Review It

Describe and differentiate the gradualistic and punctuated equilibrium models of evolution.

Name two genes which have been identified as able to create radical changes in body shapes and organs.

Use It

Which model of evolution is correct, the gradualistic or punctuated equilibrium model?

Explain how your eyes are related to the eyes of an octopus and the eyes of a fly even though they are very different in their structure and function.

Summarize It

Describe the similarities and differences between macro- and microevolution.

Does evolution have a goal?

Describe how polyploidy in plants can lead to the evolution of a new plant species.

Describe how *Hox* genes may have influenced macroevolution.

FOLLOWING *the* BIG IDEAS

AP Essential Knowledge	Chapter Section
BIG IDEA 1 **1.A.4** Biological evolution is supported by scientific evidence from many disciplines, including mathematics.	18.2
1.B.1 Organisms share many conserved core processes and features that evolved and are widely distributed among organisms today.	18.1, 18.2
1.C.1 Speciation and extinction have occurred throughout the Earth's history.	18.2
1.C.3 Populations of organisms continue to evolve.	18.3
1.D.1 There are several hypotheses about the natural origin of life on Earth, each with supporting scientific evidence.	18.1
1.D.2 Scientific evidence from many different disciplines supports models of the origin of life.	18.2

CHAPTER OVERVIEW

Life on Earth began approximately 3.5 BYA. Earth's early conditions and chemical makeup provided an environment capable of building biomolecules. There are a few different hypotheses that scientists have developed on how life arose from these molecules. The history of life on Earth is documented in the fossil record; many speciation events and extinctions have occurred.

18.1 Origin of Life

Essential Knowledge covered
1.B.1: Organisms share many conserved core processes and features that evolved and are widely distributed among organisms today.
1.D.1: There are several hypotheses about the natural origin of life on Earth, each with supporting scientific evidence.

Recall It

There are four stages in which life may have arose from biomolecules under Earth's early conditions. Briefly, organic monomers formed first, followed by organic polymers, protocells, and then, eventually, living cells. Scientists in different disciplines support the idea of these four stages, but through multiple hypotheses, such as: the "organic soup" model, the iron-sulfur world hypothesis, and the RNA-first hypothesis.

18.1 Origin of Life *continued*

Review It

Describe the different hypotheses that describe how life on Earth may have formed.

Hypothesis	Description
Abiotic Synthesis	
The Iron-Sulfur World Hypothesis	
The Membrane-First Hypothesis	
The Protein-First Hypothesis	
The RNA-First Hypothesis	
The Primordial Soup Hypothesis	

List the four stage which describe how life originated from non-living matter.

Use It

Draw a diagram to explain how the first plasma membrane might have developed. Include in your answer the words micelle, vesicle, and protocell.

Which came first: DNA, RNA, or proteins?

18.2 History of Life

Essential Knowledge covered
1.A.4: Biological evolution is supported by scientific evidence from many disciplines, including mathematics.
1.B.1: Organisms share many conserved core processes and features that evolved and are widely distributed among organisms today.
1.C.3: Populations of organisms continue to evolve.
1.D.2: Scientific evidence from many different disciplines support models of the origin of life.

Recall It

The fossil record describes the chronological order of life on Earth. Fossils can be dated absolutely using radioactive carbon dating, a radiometric technique that measures the radioactive decay of ^{14}C. Potassium-argon dating is used to date rocks 100,000 to 4.5 billion years old. Relative dating is used as well. The history of life on Earth is broken into eras, periods, and epochs. Microfossils that resemble modern day bacteria and prokaryotes are the first to appear in the fossil record. It is thought eukaryotic cells arose from nucleated cells engulfing various prokaryotes, as described by the endosymbiotic theory. The fossil record also shows that many mass extinction events have occurred throughout the history of life.

Review It

List four pieces of modern evidence that support the endosymbiotic theory.

Explain the difference between extinction and mass extinction.

Create a flow chart to describe in general terms when and how ferns evolved from algae.

Use It

Why were ancient cyanobacteria important for the evolution of other living organisms?

What is the Cambrian explosion and why do biologists find it so interesting?

18.3 Geological Factors that Influence Evolution

Essential Knowledge covered
1.C.1: Speciation and extinction have occurred throughout the Earth's history.

Recall It

The positions of the continents are not fixed and change over time in a phenomenon called continental drift. This phenomenon is studied by a branch of geology known as plate tectonics which follows the movement of pieces of the Earth's crust which float on a lower, hot mantle layer. It has been suggested that tectonics, oceanic, and climatic changes lead to mass extinctions. At least five mass extinctions have occurred in the history of life on Earth.

Review It

List the five mass extinctions that have occurred on Earth and give the probable cause of the extinction.

Use It

How did continental drift lead to differences in the distribution and diversity of marsupials between South America and Australia?

AP CHAPTER SUMMARY

Summarize It

There are several hypotheses about the natural origin of life on Earth. Pick one from the textbook. Identify the discipline it comes from, and then describe how it explains the origin of life on Earth.

What evidence, drawn from many scientific disciplines including geology and molecular biology, supports the hypothesis that all organisms on Earth share a common ancestor? What was the role of natural selection in the evolution of cells from organic precursors?

What is a mass extinction, and how does it influence speciation events?

FOLLOWING *the* BIG IDEAS

AP Essential Knowledge	Chapter Section
BIG IDEA 1 **1.A.4** Biological evolution is supported by scientific evidence from many disciplines, including mathematics.	19.3
1.B.1 Organisms share many conserved core processes and features that evolved and are widely distributed among organisms today.	19.1
1.B.2 Phylogenetic trees and cladograms are graphical representations (models) of evolutionary history that can be tested.	19.1, 19.2, 19.3

CHAPTER OVERVIEW

There are several branches of biology that allow us to explore evolutionary relationships between species. Systematic biology uses quantitative (numerically measurable) traits to infer relationships. Taxonomy classifies life into three domains—all the way to species. Cladograms are used to help construct phylogenies, by means of similar and different traits, to help trace evolutionary relationships.

19.1 Systematic Biology

Essential Knowledge covered
1.B.2: Phylogenetic trees and cladograms are graphical representations (models) of evolutionary history that can be tested.

Recall It

Identifying, naming, and organizing biodiversity is the branch of systematic biology called taxonomy. Groups are classified by sets of traits into a phylogeny. Systematic biology also uses DNA to construct phylogeny between species. Linnaean classification is used to classify species. Taxonomic groups are placed into a domain, kingdom, phylum, class, order, family, genus, and species.

Review It

List the eight main categories of biological classification.

The specific epithet of a tiger is written below.

Circle the genus and draw a square around the species.

Panthera   *tigris*

Use It

Explain how DNA can be used to identify and classify new species.

19.2 The Three-Domain System

Essential Knowledge covered
1.B.2: Phylogenetic trees and cladograms are graphical representations (models) of evolutionary history that can be tested.

Recall It

Currently, there are three domains of life: Domain Bacteria, Domain Eukarya, and Domain Archaea. Bacteria are structurally similar to Archaea, however there are distinct biochemical differences between the two. Many archaea live in extreme environments, such as in volcanic vents. Eukaryotes are single-celled to multicellular organisms but all have a membrane-bound nucleus and various organelles.

Review It

Fill in the following table to distinguish between the three domains of life.

Trait	Bacteria	Archaea	Eukarya
Single-celled			
Membrane lipids			
Ribosomes			
Cell wall			
Nuclear envelope			
Introns			

Use It

Describe how bacteria are different from archaea.

Some protists and fungi are single-celled organisms. Why are they classified as eukaryotes and not as bacteria or archaea?

19.3 Phylogeny

Essential Knowledge covered
1.A.4: Biological evolution is supported by scientific evidence from many disciplines, including mathematics.
1.B.2: Phylogenetic trees and cladograms are graphical representations (models) of evolutionary history that can be tested.

Recall It

A phylogeny is a visual representation of evolutionary history. Traits are used to build a phylogeny. Derived traits are important for clarifying evolutionary history, and ancestral traits are not as helpful. Cladograms are used to help build a phylogeny. A cladogram is a table or visual representation of derived traits of taxa being compared. Some traits are homologous; meaning they share a common ancestor; whereas some are convergent, meaning they look the same because they were under similar environmental pressures as they evolved independently. Molecular traits, behavioral traits, and even protein comparisons can all be used to build a phylogeny.

Review It

Compare and contrast homology and analogy.

Use It

If there are 3 amino acid differences in the protein cytochrome *c* between horses and pigs, 10 amino acids different between chickens and pigs, and 12 between horses and chickens, what can be concluded about the evolutionary relationships between these organisms?

Construct a cladogram, using the chart of physical traits in the organisms listed below.

Trait	Kangaroo	Human	Koala	Cat	Whale
Placental mammal					
Fins					
Four limbs					
Canine teeth					
External tail					

Summarize It

In 2006, a giant creature that looked like a sea anemone was found at the bottom of the ocean and named *Boloceroides daphneae*. If you look up this organism today you will see that is now called *Relicanthus daphneae*. Why might a scientist change an organism's name?

What types of scientific data can be used to construct phylogenetic trees or cladograms to visualize the evolutionary history of a group(s) of organisms?

UNIT THREE: EVOLUTION

Chapter 15 Darwin and Evolution • Chapter 16 How Populations Evolve • Chapter 17 Speciation and Macroevolution • Chapter 18 Origin and History of Life • Chapter 19 Taxonomy, Systematics, and Phylogeny

Multiple Choice Questions

Directions: For each question or incomplete statement chose one of the four suggested answers or completions listed below.

Question 1, 2, and 3
Scientists conducted a study on the evolution of web-building in *Tetragnatha* spiders. *Tetragnatha* spiders are endemic to the Hawaiian archipelago. Different *Tetragnatha* species are found within the forest habitats on each of the main islands and build noticeably different webs.

The following phylogenic analysis was developed for historical relationships among the species from the different islands. The illustrations above the phylogenic tree denote the three different types of webs built by the spider species. Similar web architectures are outlined in the same color.

Data obtained from: Blackledge, Todd A. and Rosemary G. Gillespie. 2004. Convergent evolution of behavior in an adaptive radiation of Hawaiian web-building spiders. Proceedings of the National Academy of Sciences 101.4: 16228-16233

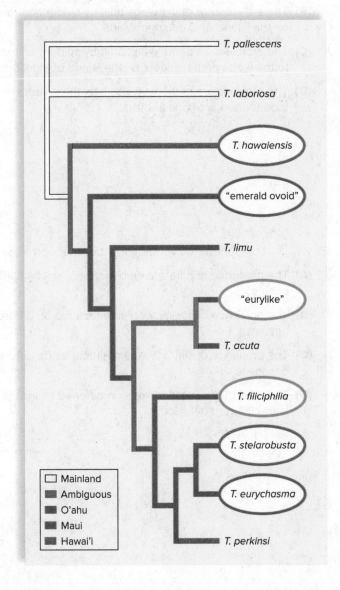

1. Darwin proposed that a small population of ancestral finches colonized the Galápagos Islands and their descendants spread out to occupy various niches. The *Tetragnatha* spiders in this study are examples of this same type of speciation, known as

   (A) adaptive radiation.

   (B) sympatric speciation.

   (C) temporal isolation.

   (D) hybrid sterility.

2. Scientists compared the web architectures of the species on and within different islands of Hawai'i. What conclusions can the scientists draw from their phylogenic analysis?

   (A) The *Tetragnatha* spiders of O'ahu and Maui are not related.

   (B) The architecture of the webs built by spiders on the island of Hawai'i are more similar to the webs on the island of O'ahu than Maui.

   (C) The architecture of the webs built by spiders on the island the island of Hawai'i are more similar to the webs of the spiders on the island of Maui than O'ahu.

   (D) The architecture of the webs built by the spiders on different islands most likely arose independently of one another.

3. Which of the following hypotheses is best supported by these data?

   (A) The presence of different web architectures by spiders on similar islands is not a product of evolution.

   (B) The presence of similar web architectures on different islands suggests convergent evolution occurred.

   (C) The presence of similar web architectures on different islands suggests divergent evolution occurred.

   (D) The presence of similar webs on different islands suggests forests on the different Hawaii islands have different habitats.

Questions 4 and 5

The following cladogram shows the evolution of changes in swimming mode during whale evolution:

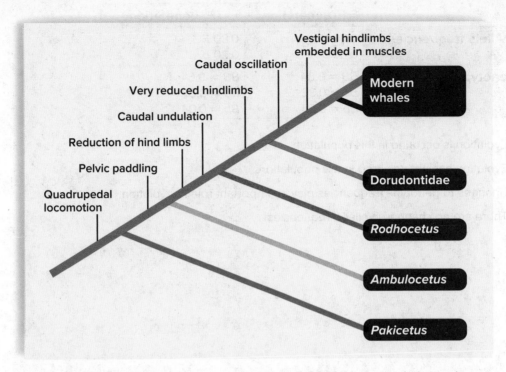

Data obtained from: Bejder, Lars, and Brian K. Hall. 2002. Limbs in whales and limblessness in other vertebrates: mechanisms of evolutionary and developmental transformation and loss. *Evolution & development* 4.6: 445-458

4. Which best describes the evolution of change in swimming mode during whale evolution according to this cladogram?

   (A) Quadrupedal locomotion is a trait derived due to evolution.

   (B) Quadrupedal locomotion is a trait found in modern whales.

   (C) Quadrupedal locomotion is a trait lost due to evolution.

   (D) Quadrupedal locomotion is a trait derived from pelvic paddling.

5. Modern whales are most closely related to which of the following species?

   (A) *Pakicetus*

   (B) *Ambulocteus*

   (C) *Rodhocetus*

   (D) All of the above

6. The following table represents the change in genotype and phenotype frequencies in a given population. What conclusions can you draw about the evolution of this population?

	F$_1$ generation	F$_2$ generation
Allele frequencies	0.20 B 0.80 b	0.80 B 0.20 b
Genotype frequencies	BB = 0.04 Bb = 0.32 Bb = 0.64	BB = 0.64 Bb = 0.32 Bb = 0.04

(A) Evolution is occuring in this population.

(B) Evolution has not occurred in this population.

(C) Changes to genotype frequencies play an important role in evolution.

(D) There are no changes in allelic frequencies.

Free Response Questions

Directions: Read the questions carefully and completely. Then, plan your answer and write your response in the space provide. Write your answer out in paragraph form.

1. A population of tree frogs have been found living on a remote island. There are two larger islands, one to the west and one to the east. Similar tree frogs inhabit these islands as well, which look similar but have slight differences in their foraging behaviors and diets.

   (a) Describe TWO types of data that could be collected to determine if the tree frogs on the remote island are related to either of the tree frog populations found on the islands to the east and west.

   (b) If the data collected indicated that these frogs are three different species, although still closely related to each other, what might you conclude about the evolutionary differences in these organisms?

2. The following graph shows the relationship between human birthweight and infant mortality.

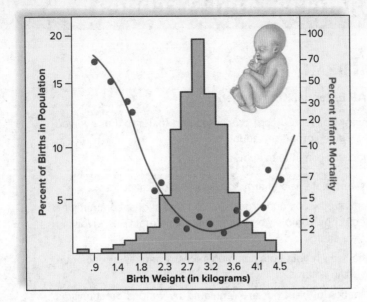

(a) Describe what type of natural selection may be acting on this population, and explain how you came to your conclusion.

(b) Provide an explanation on how this leads to a reduction of variability in the birthweight of humans.

FOLLOWING *the* BIG IDEAS

AP Essential Knowledge	Chapter Section
BIG IDEA 1 **1.B.1** Organisms share many conserved core processes and features that evolved and are widely distributed among organisms today.	20.2
1.C.3 Populations of organisms continue to evolve.	20.1
BIG IDEA 2 **2.A.2** Organisms capture and store free energy for use in biological processes.	20.4
2.D.1 All biological systems from cells and organisms to populations, communities and ecosystems are affected by complex biotic and abiotic interactions involving exchange of matter and free energy.	20.3
2.D.2 Homeostatic mechanisms reflect both common ancestry and divergence due to adaptation in different environments.	20.2
2.E.3 Timing and coordination of behavior are regulated by various mechanisms and are important in natural selection.	20.3
BIG IDEA 3 **3.A.1** DNA, and in some cases RNA, is the primary source of heritable information.	20.1
3.C.2 Biological systems have multiple processes that increase genetic variation.	20.2
3.C.3 Viral replication results in genetic variation, and viral infection can introduce genetic variation into the hosts.	20.1, 20.3
3.D.4 Changes in signal transduction pathways can alter cellular response.	20.3
BIG IDEA 4 **4.B.2** Cooperative interactions within organisms promote efficiency in the use of energy and matter.	20.3, 20.4

CHAPTER OVERVIEW

Viruses, bacteria, and archaea may be quite small, but they are major players on Earth. Viruses are obligate intracellular parasites that have DNA or RNA, and use the machinery of a host cell to replicate. Prokaryotes include bacteria and archaea. These single-celled organisms occupy every niche imaginable on Earth and have diverse methods of gathering and metabolizing energy.

20.1 Viruses, Viroids, and Prions

Essential Knowledge covered
1.C.3: Populations of organisms continue to evolve.
3.A.1: DNA, and in some cases RNA, is the primary source of heritable information.
3.C.3: Viral replication results in genetic variation, and viral infection can introduce genetic variation to the hosts.

Recall It

A virus is known as an obligate intercellular parasite. Viruses are composed of a capsid and an inner core of nucleic acid. Viruses that invade bacteria are known as bacteriophages and alternate between two life cycles: the lysogenic cycle and the lytic cycle. Lysogenic cells carry prophage genes. In animals, it is possible that viruses can carry RNA, convert it into DNA, and integrate it into the host's genome with the help of the enzyme reverse transcriptase. These viruses are known as retroviruses. New viruses that infect large numbers of humans are called emerging viruses and are particularly difficult to cure due to their rapid evolution. While not technically viruses because they lack capsids, viroids are able to cause disease in plants. Infectious protein particles called prions can cause neurodegenerative diseases by forming clusters in the brain.

20.1 Viruses, Viroids, and Prions *continued*

Review It

List the five steps of the reproductive cycle of viruses.

Using a labeled diagram, explain how some RNA viruses use reverse transcription in order to be integrated into the host genome to produce more viruses.

Use It

Describe why H5N1 and H7N9 are considered emerging viruses.

How can a virus cause the bacterium that causes strep throat cell to produce a toxin that causes scarlet fever?

20.2 The Prokaryotes

Essential Knowledge covered
1.B.1: Organisms share many conserved core processes and features that evolved and are widely distributed among organisms today.
2.D.2: Homeostatic mechanisms reflect both common ancestry and divergence due to adaptation in different environments.
3.C.2: Biological systems have multiple process that increase genetic variation.
3.C.3: Viral replication results in genetic variation, and viral infection can introduce genetic variation to the hosts.

Recall It

Prokaryotes are fully-functioning, living, single-celled organisms. Prokaryotes are found in two domains: the Bacteria and the Archaea. Prokaryotes lack a nucleus and organelles. Bacteria and Archaea inhabit many different habitats and utilize many different energy sources. Prokaryotes reproduce asexually by binary fission. Conjugation, transformation, and transduction are all means of genetic recombination that have been observed in prokaryotes.

20.2 The Prokaryotes *continued*

Review It

Given the definition on the left, fill in the correct term on the right:

Definition	Vocabulary Word
When a cell picks up free pieces of DNA secreted or released by a prokaryote	
When two bacteria link together and one passes DNA to the other in the form of a plasmid	
Fully-functioning, living, single-celled organisms	
Asexual reproduction in prokaryotes	
Process in which bacteriophages carry portions of DNA from one bacterial cell to another	
The elongated, hollow appendage used to transfer DNA between bacterial cells	

List the three types of appendages a prokaryote might have.

Use It

List three processes prokaryotes can carry out which allow for the acquisition and recombination of new genetic information.

Why are the three processes you listed above important for achieving genetic variation in prokaryotes?

Compare and contrast transduction and transformation using the diagram below:

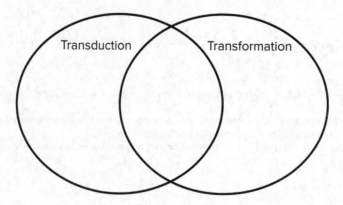

20.2 The Prokaryotes *continued*

Describe two ways prokaryotes have adapted to a wide range of environments.

20.3 The Bacteria

Essential Knowledge covered
2.D.1: All biological systems from cells and organisms to populations, communities and ecosystems are affected by complex biotic and abiotic interactions involving exchange of matter and free energy.
2.E.3: Timing and coordination of behavior are regulated by various mechanisms and are important in natural selection.
3.D.4: Changes in signal transduction pathways can alter cellular response.
4.B.2: Cooperative interactions within organisms promote efficiency in the use of energy and matter.

Recall It

Bacteria are prokaryotic organisms that are found almost ubiquitously on Earth. The more common type of prokaryote, bacteria are protected by a cell wall composed of peptidoglycan. When faced with unfavorable conditions, some bacteria can release a portion of their cytoplasm and a copy of their chromosome in an endospore. Some bacteria can produce a toxin to damage other organisms. There are many types of metabolic activities found within the bacteria. Some bacteria are chemoautotrophs, chemoheterotrophs, photoautotrophs, or saprotrophs. Many bacteria are found in symbiotic relationships with other organisms.

Review It

Identify the bacteria based on the description of its metabolic activity.

Metabolism	Bacteria
Autotrophic bacteria which carry out chemosynthesis	
Heterotrophic bacteria which obtain organic nutrients from other living organisms	
Autotrophic bacteria which are photosynthetic	
Heterotrophic bacteria which break down organic matter from dead organisms	
Gram-negative bacteria that photosynthesize and may contain other pigments as well	

Identify if the relationship is either parasitic (P), commensal (C), or mutualistic (M).

An obligate anaerobe lives in our gut because *E. coli* has used up the oxygen.

A prokaryote digests cellulose in the gut of a goat, and the goat eats grass.

The bacterium, *Clostridium tetani*, enters a cut on someone's foot and colonizes the site of the wound.

A cyanobacterium resides in the nodule of a legume root and the plant is able to obtain nitrogen from it.

A pathogen, *Shigella dysenteriae*, sticks to the human intestinal wall and produces Shiga toxin.

20.3 The Bacteria *continued*

Use It

Describe a mutualistic relationship between bacteria and a eukaryote. Use a specific example you found interesting in the text. Be sure to identify why it's considered mutualistic.

Why do some *Salmonella* strains make us sicker than others?

20.4 The Archaea

Essential Knowledge covered
2.A.2: Organisms capture and store free energy for use in biological processes.
2.D.2: Homeostatic mechanisms reflect both common ancestry and divergence due to adaptation in different environments.
4.B.2: Cooperative interactions within organisms promote efficiency in the use of energy and matter.

Recall It

Archaea are prokaryotic organisms that are similar to bacteria but are different in their biochemical makeup. Many archaea can live under especially harsh conditions. Three main types of archaea include: methanogens, halophiles, and thermoacidophiles. Archaea can also be found in more moderate habitats and in symbiotic relationships with animals as well.

Review It

Fill in the chart to describe the three types of archaea.

Name	Habitat	Energy Capturing Mechanism	Unique Trait
Methanogens			
Halophiles			
Thermoacidiophiles			

List three reasons why archaea are more similar to eukaryotes then bacteria.

How have halophiles adapted to living in saline environments?

20.4 The Archaea *continued*

Use It

What features allow many archaea to function at high temperatures?

A hydrothermal vent is located in a deep-ocean vent. Here no sunlight shines, the water can reach up to 140°C, and is rich in acidic hydrogen sulfide gas. What type of archaea could a scientist expect to find here and why?

Around the vent there is a complex biological community thriving with macroinvertebrates, fish, and octopuses. How do these archaea help in building this community?

AP CHAPTER SUMMARY

Summarize It

The fossil record suggests that prokaryotes were alone on Earth for 2.5 billion years. How does this information help us better understand the diversity we see in prokaryotes today?

Suppose the water in a lake near a big farm suddenly becomes cloudy and green in the middle of the summer. A scientist on a local news channel says that it is full of cyanobacteria because of high levels of fertilizer present in the area. What does this mean? What does it mean for the future of the lake?

Describe how a nitrifying marine archaea may impact other organisms living in the ocean.

FOLLOWING *the* BIG IDEAS

AP Essential Knowledge	Chapter Section
BIG IDEA 1 **1.B.1** Organisms share many conserved core processes and features that evolved and are widely distributed among organisms today.	21.1
2.D.1 All biological systems from cells and organisms to populations, communities and ecosystems are affected by complex biotic and abiotic interactions involving exchange of matter and free energy.	21.1
2.D.2 Homeostatic mechanisms reflect both common ancestry and divergence due to adaptation in different environments.	21.1
BIG IDEA 4 **4.B.4** Distribution of local and global ecosystems change over time.	21.1
4.C.3 The level of variation in a population affects population dynamics.	21.1

CHAPTER OVERVIEW

Protists are the simplest, but most diverse branch of eukaryotes. Protists are classified into six supergroups, each with a different evolutionary lineage. Some of the protists discussed in this chapter include: algae, water molds, zooflagellates, protozoans, radiolarians, and forminiferans.

21.1 General Biology of Protists

Essential Knowledge covered
2.A.2: Organisms capture and store free energy for use in biological processes.
2.D.2: Homeostatic mechanisms reflect both common ancestry and divergence due to adaptation in different environments.

Recall It

Protists are eukaryotic organisms that contain membranous organelles. They are often grouped based on how they collect organic nutrients. Some are heterotrophic, some are autotrophic, and some are mixotrophic—meaning they can utilize both autotrophic and heterotrophic methods to acquire nutrients. There are currently six supergroups of protists: (1) Archaeplastida, (2) Chromalveolata, (3) Excavata, (4) Amoebozoa, (5) Opisthokonta, and (6) Rhizaria. These multiple protist lineages arose separately according to DNA evidence.

Review It

How do protists reproduce?

How are protists categorized?

21.1 General Biology of Protists *continued*

Use It

Describe two ways in which protists play a significant ecological role in aquatic systems.

As a whole, protists are very diverse; however, many share similar characteristics. How might you explain both different and similar traits within protists?

21.2 Supergroup Archaeplastida
Extending Knowledge

Recall It

The details of the protist supergroups are beyond the scope of the AP exam. Illustrative examples of Big Ideas 1 and 2 can be found in this section, such as the conserved processes between green algae and land plants.

Review It

Describe the characteristic all members of the supergroup Archaeplastida have in common.

Describe the similarities in structure and ecological role between green algae and land plants.

21.3 Supergroup Chromalveolata
Extending Knowledge

Recall It

The details of the protist supergroups are beyond the scope of the AP exam. Illustrative examples of Big Ideas 2 and 4 can be found in this section, such as the ecological factors that lead to red tides and the lifecycle of pathogen which causes malaria.

21.3 Supergroup Chromalveolata *continued*
Extending Knowledge

Review It

Kelp and *Fucus* are brown algae that live along the shoreline in temperate regions. What sorts of mechanisms have they evolved to be able to withstand both powerful waves and exposure to air?

What are red tides, and how can they be dangerous?

Describe what causes malaria and how it is spread through a population.

21.4 Supergroup Excavata
Extending Knowledge

Recall It

The details of the protist supergroups are beyond the scope of the AP exam. Illustrative examples of Big Ideas 2 and 4 can be found in this section, such as the diverse trophic adaptations of euglenids and the lifecycle of the pathogen which causes African Sleeping Sickness.

Review It

Why is it difficult to classify the Euglenids?

Draw a diagram of how African Sleeping Sickness is transmitted.

21.5 Supergroups Amoebozoa, Opisthokonta, and Rhizaria
Extending Knowledge

Recall It
The details of the protist supergroups are beyond the scope of the AP exam. Illustrative examples of Big Ideas 1 and 2 can be found in this section, such as homeostatic mechanisms employed by amoebas and the role that foraminiferans play in nutrient cycling.

Review It
Describe how amoebozoans move.

How do foraminiferans help shape the environment?

AP CHAPTER SUMMARY

Summarize It
What explains the amazing diversity of protists, and why is it challenging to classify them?

Describe a way in which climate change may increase the rate of protist-related human disease.

FOLLOWING *the* BIG IDEAS

	AP Essential Knowledge	Chapter Section
BIG IDEA 1	**1.B.1** Organisms share many conserved core processes and features that evolved and are widely distributed among organisms today.	22.1
BIG IDEA 2	**2.C.2** Organisms respond to changes in their external environment.	22.1, 22.2
	2.E.2 Timing and coordination of physiological events are regulated by multiple mechanisms.	22.2, 22.3
BIG IDEA 4	**4.B.3** Interactions between and within populations influence patterns of species distribution and abundance.	22.3
	4.C.3 The level of variation in a population affects population dynamics.	22.3

CHAPTER OVERVIEW

22.1 Evolution and Characteristics of Fungi

Essential Knowledge covered
1.B.1: Organisms share many conserved core processes and features that evolved and are widely distributed among organisms today.
2.C.2: Organisms respond to changes in their external environment.

Recall It

Fungi are heterotrophs that are spilt into six groups. As they obtain their nutrients by absorbing food, fungi are known as saprotrophs. Some fungi produce filaments called hyphae which group from their tips and may have cross-walls known as septa. Septate fungi have these cross-walls and aseptate fungi do not. Dikaryotic fungi have hypha that contain paired haploid nuclei. Hyphae may form a network called mycelium. Fungi cells are different from plant cells because they contain chitin. Fungi usually produce spores during sexual and asexual reproduction. Other fungi can reproduce asexually by budding.

Review It

How are fungi able to disperse their offspring?

Describe two ways fungi are more like animals than plants.

22.1 Evolution and Characteristics of Fungi *continued*

Use It

Draw a picture showing the three stages of sexual reproduction in terrestrial fungi.

A terrestrial fungi needs to obtain nutrients but its food source is half a kilometer away. Since the fungi is non-motile, how would it reach its food?

22.2 Diversity of Fungi
Extending Knowledge

Recall It

Fungi, like protists, have complex phylogenetic relationships. Currently, fthere are six groups of fungi: Chytridiomycota, Zygomycota, Ascomycota, Basidiomycota, Glomeromycota, and Microsporidia. The details of these groups are outside the scope of the AP exam, though you will find several applications of the AP Big Ideas 2 and 4 in the study of fungi diversity.

Review It

Draw a picture of the life cycle of black bread mold. When are the cells haploid? When are they diploid?

How does black bread mold survive when conditions are unfavorable?

What type of fungi do leaf cutter ants and elm trees have in common?

22.3 Symbiotic Relationships of Fungi

Essential Knowledge covered
2.E.2: Timing and coordination of physiological events are regulated by multiple mechanisms.

Recall It

There are several types of fungi that have formed symbiotic relationships with other organisms. Lichens are an association between a fungus and cyanobacterium. Lichens are found growing on trees or rocks and can be classified by their shapes. The symbiotic relationship is now thought be parasitic, as the algae may not benefit from the relationship. Mycorrhizae are mutualistic relationships between soil fungi and roots of most plants. Mycorrhizae allow for increased surface area on roots to better absorb more nutrients and water from the soil. In return, the fungus receives carbohydrates from the plant.

Review It

Describe the three types of lichen.

What is the difference between ectomycorrhizae and endomycorrhizae?

Use It

Why is the formation of lichen no longer considered mutualistic?

Describe a benefit received by the plant and by the fungus in a mycorrhizae relationship.

AP CHAPTER SUMMARY

Summarize It

Describe how fungi are related to plants and animals.

Explain what would happen to other organisms on Earth if all fungi suddenly went extinct.

Describe two ways fungi adapt to changing environmental conditions.

UNIT FOUR: MICROBIOLOGY AND EVOLUTION

Chapter 20 Viruses, Bacteria, and Archaea ● Chapter 21 Protist Evolution and Diversity
● Chapter 22 Fungi Evolution and Diversity

Multiple Choice Questions

Directions: For each question or incomplete statement choose one of the four suggested answers or completions listed below.

1. An emerging virus was found in a population of monkeys. Viral infections increased exponentially over a period of ten years. Given this information, what statement best describes the nature of this virus?

   (A) This virus enters directly into the lytic cycle.

   (B) This virus enters directly into the lysogenic cycle.

   (C) This virus is highly virulent over a short period of time.

   (D) This virus is bacterial in nature.

Use the information below for questions 2 and 3.

Aster yellows is a plant disease causes a number of different abnormalities in several species of plants, such as: stunted plant growth, loss of pigmentation, and sterility in flowers. Plants with this disease first experience predation by leafhoppers – insects which feed on the leaves or stems of the plant.

A series of scientific studies has provided the following data on Aster yellows.

   1. Repeated electron microscopic examinations of leaves and shoots of infected plants have shown organisms that look most like *Mycoplasma*, a genus of bacteria that lack a cell wall around their cell membrane.
   2. These organisms have also been found in the salivary glands of leafhopper
   3. These organisms are also found in the roots and flowers of infected plants but not found in healthy plants.

2. Using this provided information, which of the following statements can be made about Aster yellows:

   (A) Aster yellows is caused by leaf hoppers.

   (B) Aster yellows is caused by a virus.

   (C) Aster yellows is caused by a prokaryote.

   (D) Aster yellows is caused by eukaryotes.

3. Further research shows that the organism inside the infected plants and leaf hoppers is unable to be cultured and survive on its own. The relationship between the organism and the plant is most likely

(A) commensal.

(B) nitrogen-fixing.

(C) parasitic.

(D) mutualistic.

4. Which evolutionary trait is most advantageous for plasmodial slime molds to survive adverse environmental conditions, such as a month-long drought?

(A) The ability to creep along the forest floor as a mass of flagellated cells.

(B) The ability to consume decaying plant material.

(C) The ability to produce spores.

(D) The ability to produce amoeboid cells.

5. Some protists use contractile vacuoles to remove excess water from their cells. If the function of this vacuole is diminished, what will happen to a protist cell?

(A) The cell will swell until it ruptures.

(B) The cell will shrivel up.

(C) The cell will remain in an isotonic state.

(D) The cell will not change.

6. Chytrids are a unique single-celled species of fungi that reproduce by creating zoospores. Some of these zoospores grow into new chytrids via asexual reproduction. However, some have an alternation-of-generation life cycle. This means

(A) haploid gametes can combine to produce a diploid zygote.

(B) haploid gametes can combine to produce a haploid zygote.

(C) diploid gametes can combine to produce diploid zygotes.

(D) diploid gametes can combine to produce haploid zygotes.

Free Response Questions

Directions: Read the questions carefully and completely. Then, plan your answer and write your response in the space provide. Write your answer out in paragraph form.

1. Sloths are slow-moving, arboreal mammals. They have highly absorbent hair, which has been observed to look green at times. A study took a closer look at what causes this change in pigmentation, and discovered a host of different organisms living inside sloth hair.

Composition of Organisms in Sloth Hair	
Algae	27%
Ciliates, apicomplexans and dinoflagellates	52%
Fungi	8%
Others	13%

(a) Looking at this dataset, provide an explanation as to why sloth hair sometimes looks green.

(b) Describe the different types of species found in the sloth hair. What characteristics do these organisms share?

(c) Suggest a possible hypothesis as to why sloths have these different organisms growing in their hair.

2. Zika virus is an emerging virus which can cause microcephly (abnormally small head and brain size) in infants born to infected mothers. The Zika virus has an envelope and a single RNA molecule. Zika virus enters cells by endocytosis, and it is seen in the cytoplasm and in the nucleus of cells.

Explain how Zika is able to enter a human cell, replicate, and cause physiological changes in a fetus, if it's only a single strand of RNA.

23 Plant Evolution and Diversity

FOLLOWING *the* BIG IDEAS

AP Essential Knowledge	Chapter Section
BIG IDEA 1 **1.A.1** Natural selection is a major mechanism of evolution.	23.1, 23.3, 23.4
1.B.1 Organisms share many conserved core processes and features that evolved and are widely distributed among organisms today.	23.4
1.B.2 Phylogenetic trees and cladograms are graphical representations (models) of evolutionary history that can be tested.	23.1, 23.2, 23.4

CHAPTER OVERVIEW

From their aquatic ancestors, plants have evolved to become extremely diverse and carry out critical processes that support life on Earth. There are four major groups of land plants: (1) mosses, (2) ferns, (3) gymnosperms, and (4) angiosperms. Each group has distinguishing characteristics well suited for life on land.

23.1 Ancestry and Features of Land Plants

Essential Knowledge covered
1.A.1: Natural selection is a major mechanism of evolution.
1.B.2: Phylogenetic trees and cladograms are graphical representations (models) of evolutionary history that can be tested.

Recall It

Plants are thought to have evolved from single-celled aquatic alga into multicellular, photosynthetic eukaryotes. Plants have many different features which allowed them to adapt to land to prevent desiccation. All land plants have an alternation of generations life cycle.

Review It

List two ways green algae are like plants and one major difference.

According to phylogenetic information, in what order did modern plant groups diverge?

23.1 Ancestry and Features of Land Plants *continued*

Use It

Describe four adaptations that allowed plants to move to land.

Draw and explain a diagram of the alternation of generations in land plants. Label the diploid and haploid structures as well when meiosis and mitosis occur.

23.2 Evolution of Bryophytes: Colonization of Land

Essential Knowledge covered
1.B.2: Phylogenetic trees and cladograms are graphical representations (models) of evolutionary history that can be tested.

Recall It

Bryophytes include mosses, liverworts, and hornworts, and are thought to be the first land plants to evolve. As such, they have rudimentary stems, roots, and leaves which lack vascular tissue. Therefore, they are found in moist environments where they still have lots of access to water.

Review It

Describe two defining characteristics for each bryophyte in the table.

Moss	Liverwort	Hornwort

Use It

Compare and contrast traits of bryophytes with traits of vascular plants.

23.3 Evolution of Lycophytes: Vascular Tissue

Essential Knowledge covered
1.A.1: Natural selection is a major mechanism of evolution.
1.B.2: Phylogenetic trees and cladograms are graphical representations (models) of evolutionary history that can be tested.

Recall It

Lycophytes, also known as the club mosses, are organized into three main groups: the ground pines, spike mosses, and quillworts. Lycophytes represent an important evolutionary leap for plants as they were the first to have vascular tissue called xylem with lignin in the cell walls. Xylem is fundamental for plants to grow upright and taller.

Review It

List two ways the branching of *Cooksonia* supports important information concerning the evolution of vascular plants.

Use It

How was xylem essential to the evolution of upright and taller plants?

23.4 Evolution of Pteridophytes: Megaphylls

Essential Knowledge covered
1.A.1: Natural selection is a major mechanism of evolution.
1.B.1: Organisms share many conserved core processes and features that evolved and are widely distributed among organisms today.
1.B.2: Phylogenetic trees and cladograms are graphical representations (models) of evolutionary history that can be tested.

Recall It

Pteridophytes include ferns, horsetails, and whisk ferns. Pteridophytes are seedless and reproduce using spores, but have megaphylls; broad leaves with vascular tissue.

Review It

Identify the pteridophyte based on its description.

Description	Pteridophyte
Members of the genera *Psilotum* or *Tmesipteris*	
All members of the genus *Equisetum*	
The most diverse group of pteridophytes that range in size from 1 cm in diameter to 20 m high	

Describe the difference between megaphyll and a microphyll.

23.4 Evolution of Pteridophytes: Megaphylls *continued*

Use It

List two ways the megaphylls of ferns are important to the evolution of vascular plants.

23.5 Evolution of Seed Plants: Full Adaptation to Land
Extending Knowledge

Recall It

Seed plants are vascular plants that produce a sporophyte embryo, that has a protective coat and contains nutrients. There are two groups of seed plants: gymnosperms and angiosperms. Gymnosperms have ovules that are not enclosed by sporophyte tissue at the time of pollination; whereas, angiosperms have ovules that are completely enclosed within diploid sporophyte tissue. Gynmosperms include the conifers, cycads, gingkgoes, and gnetophytes. Angiosperms are exceptionally diverse.

Review It

Determine if the following statement applies to angiosperms, gymnosperms or both.

Statement	Type of Seed Plant
Can be monocot or eudicot	
Contains conifers, cycads, ginkgoes, and gnetophytes	
Has seeds	
Ovules are completely enclosed and become fruit	
May be monoecious or dioceious	
Ovules are not completely enclosed by sporophyte tissue	

Why was pollen an important evolutionary advantage?

Summarize It

What environmental challenges did plants face in their evolution from aquatic to terrestrial environments? What adaptations enabled plants to make this transition?

Describe the overarching evolution of land plants. What characteristics do all land plants share and which traits evolved over time to make plants more successful?

FOLLOWING *the* BIG IDEAS

AP Extending Knowledge

The material in Chapter 24 is largely outside the scope of the AP exam, however, the details on the structure of flowering plants serve as illustrative examples for Big Ideas 2 and 4.

CHAPTER OVERVIEW

Angiosperms, also known as flowering plants, have specialized cells that carry out specialized functions. These cells are organized into particular tissues and organs which play important roles in protection, metabolism, and reproduction. Xylem and phloem are tissues critical in the transport of water and nutrients, respectively. Flowers are organs unique to angiosperms which are key to sexual reproduction. Angiosperms are used by humans in numerous ways, from foods to medicines, culture, and cosmetics.

24.1 Cells and Tissues of Flowering Plants
Extending Knowledge

Recall It

Flowering plants develop from meristems into apical meristems, and then divide and differentiate into differing plant tissues. Flowering plants have many types of specialized tissue. Epidermal tissue has a waxy cuticle. Root epidermal cells have root hairs, and epidermal cells of stems, leaves, and flowers can produce trichomes. Stomata, openings in the cell, play a role in gas exchange and water loss. The most tissue of a flowering plant is ground tissue, which contains parenchyma, collenchyma, and sclerenchyma. The vascular tissues xylem and phloem transport water and minerals from the roots to the leaves respectively.

Review It

Identify the structure given its function.

Function	Structure
Minimizes water loss and protects against disease	
Transports sucrose and organic compounds from the leaves to the roots	
Open and close for gas exchange and water loss	
Cells which give flexible support to immature regions of the body	
Transports water and minerals from the root to the leaves	
Can assume many different shapes and sizes as they grow	
The cork area of a plant which helps a plant resist predation	
The most abundant cells in a plant which may contain chloroplasts or may just store carbohydrates	
Thick secondary cell walls impregnated with lignin that make cells strong and hard	
Hairs which protect the plant from sun and moisture loss	
Increases the surface area for nutrient and water absorption in roots	

24.1 Cells and Tissues of Flowering Plants *continued*
Extending Knowledge

Describe how xylem moves water from the roots to the leaves.

Describe how phloem moves sucrose from the leaves to the roots.

24.2 Organs of Flowering Plants
Extending Knowledge

Recall It

Angiosperms contain vegetative and reproductive organs. The root, the stem, and the leaf make up the vegetative organs; whereas, the flower and pollen structures would be classified as reproductive organs. Plants with seeds are divided into two groups: (1) monocots and (2) eudicots. This classification is based on the number of seed leaves or cotyledons in the embryonic plant. The vascular systems are also organized differently in monocots versus eudicots. Plant life cycles can be annual, living for only one season, or perennial – existing or three or more, which is determined by two flower-inducing genes.

Review It

Draw a picture of how a basic plant body is organized. Be sure to identify the root and shoot system. Include in your diagram a terminal bud, stem, node, internode, leaf, and root.

How do the root and shoot system in plants interact?

24.3 Organization and Diversity of Roots
Extending Knowledge

Recall It

Roots are broken into three different zones: (1) the zone of maturation, (2) the zone of elongation, and (3) the zone of cell division. Various specialized cells are found fully differentiated in the zone of maturation. Epidermis, cortex, endodermis, pericycle, and vascular tissue are types of specialized tissue found in roots. They serve many roles, such as protection, food storage, and regulation of water and nutrient uptake and transport. Roots tend to either be a taproot or a fibrous root, and many plants have specialized root nodules which hold nitrogen-fixing symbiotic bacteria.

24.3 Organization and Diversity of Roots *continued*
Extending Knowledge

Review It

Define the specialized tissues found in the zone of maturation and place them in order from outermost layer (1) to innermost layer (5).

Tissue	Definition	Order
Cortex		
Pericycle		
Vascular tissue		
Endodermis		
Epidermis		

Compare and contrast taproots and fibrous roots.

Describe two specialized root adaptations found in different plants.

24.4 Organization and Diversity of Stems
Extending Knowledge

Recall It

Stems carry leaves, flowers, and supports a plant's weight. Many angiosperms have herbaceous stems that exhibit mostly primary growth. Other plants have woody stems, which have both primary and secondary growth as well as a vascular cambium. Bark on a tree contains cork, cork cambium, cortex, and phloem. While it takes more energy to produce secondary tissue, woody plants have a greater defense mechanism and longer life cycles than herbaceous plants.

24.4 Organization and Diversity of Stems *continued*
Extending Knowledge

Review It

Identify the location of the annual ring, cork, cork cambium, vascular cambium, and wood.

24.5 Organization and Diversity of Leaves
Extending Knowledge

Recall It

The majority of photosynthesis is conducted in the leaves of a plant. Leaves contain specialized tissues called palisade mesophyll and spongy mesophyll. Leaves can be arranged in many different ways upon a stem, and the blade of a leaf can also be arranged in different manners, leading to much diversity in leaves. Further modifications, such as spines, tendrils, and extremely waxy epidermal coverings, increases the diversity of leaves.

Review It

While leaves can be incredibly diverse in color, shape, size, and placement on a stem, what are some conserved characteristics which all leaves share?

How have the Venus fly trap and sundew plants evolved to live in nitrogen-lacking soils?

AP CHAPTER SUMMARY

Summarize It

What structural and physiological features of angiosperms have enabled them to become the dominant form of plant life on Earth today?

25 Flowering Plants: Nutrition and Transport

FOLLOWING *the* BIG IDEAS

AP Essential Knowledge	Chapter Section
BIG IDEA 2 **2.A.3** Organisms must exchange matter with the environment to grow, reproduce and maintain organization.	25.1
2.B.2 Growth and dynamic homeostasis are maintained by the constant movement of molecules across membranes.	25.1
2.C.1 Organisms use feedback mechanisms to maintain their internal environments and respond to external environmental changes.	25.3
2.D.1 All biological systems from cells and organisms to populations, communities and ecosystems are affected by complex biotic and abiotic interactions involving exchanges of matter and free energy.	25.2
2.E.2 Timing and coordination of physiological events are regulated by multiple mechanisms.	25.3
2.E.3 Timing and coordination of behavior are regulated by various mechanisms and are important in natural selection.	25.2, 25.3
BIG IDEA 4 **4.B.2** Cooperative interactions within organisms promote efficiency in the use of energy and matter.	25.2, 25.3

CHAPTER OVERVIEW

Plants use water to conduct macro- and micronutrients from their roots to their leaves. As most plants grow in soil, soil type and composition plays a critical role in plant growth. Water moves into a plant passively through xylem; whereas, nutrients are taken up through both passive and active transport mechanisms in phloem. The special properties of water covered in Chapter 2; water potential, cohesion, and adhesion, are critical in order for tall plants to perform transpiration.

25.1 Plant Nutrition and Soil

Essential Knowledge covered
2.A.3: Organisms must exchange matter with the environment to grow, reproduce, and maintain organization.
2.B.2: Growth and dynamic homeostasis are maintained by the constant movement of molecules across membranes.

Recall It

Plants require many different types of nutrients. These are broken into categories known as macro- and micronutrients, as well as beneficial nutrients. Nitrogen (N), phosphorus (P), and potassium (K) are key macronutrients found in most fertilizers. Nutrients are acquired from the soil. Mineral ions enter a plant through the roots through a process called cation exchange. The property of soil a plant grows in is very important to the health of the plant. Healthy soil has a variety of soil particle sizes and a mix of decaying organic material, water, minerals, and air.

Review It

What makes a nutrient essential?

25.1 Plant Nutrition and Soil *continued*

Describe the difference between soil profile and soil horizon.

Use It

A ratio of humus and clay is important to support healthy plant life in soil. Why is this?

Identify two ways soil erosion causes problems.

25.2 Water and Mineral Uptake

Essential Knowledge covered
2.D.1: *All biological systems from cells and organisms to populations, communities and ecosystems are affected by complex*
biotic and abiotic interactions involving exchange of matter and free energy.
2.E.3: *Timing and coordination of behavior are regulated by various mechanisms and are important in natural selection.*
4.B.3: *Interactions between and within populations influence patterns of species distribution and abundance.*

Recall It

Water enters a root cell when the osmotic pressure in the root tissue is lower than that of the soil solution. Minerals, on the other hand, can be taken up both actively or passively. Water and minerals can travel via porous cell walls but then must enter endodermal cells because of the Casparian strip. Alternatively, water and minerals can enter root hairs and move from cell to cell. Minerals can cross an endodermal plasma membrane by way of an ATP-driven pump that establishes an electrochemical gradient that allows positively charged ions to cross the membrane via a channel protein. Negatively charged mineral ions can cross the membrane by way of a carrier when they co-transport with hydrogen ions, which diffuse down their concentration gradient. Mutualistic, symbiotic relationships with bacteria or fungi help plants aid in increased nutrient uptake.

Review It

List the two ways water can enter a root.

Determine if the following statements about mycorrhizae are true or false (T or F).

Mycorrhizae are bacteria which live on plant roots.

Most all plants have mycorrhiza.

Mycorrhizae increase the surface area for mineral and water uptake.

Plants do not depend on mycorrhizae.

Compare and contrast mycorrhizae and rhizobia.

25.2 Water and Mineral Uptake *continued*

Use It

Plants do not always obtain nutrients through mutualistic means. Describe two adaptations in plants that have evolved to obtain nutrients through other means.

25.3 Transport Mechanisms in Plants

Essential Knowledge covered
1.B.1: Organisms share many conserved core processes and features that evolved and are widely distributed among organisms today.
2.C.1: Organisms use feedback mechanisms to maintain their internal environments and respond to external environmental change.
2.E.2: Timing and coordination of physiological events are regulated by multiple mechanisms.
2.E.3: Timing and coordination of behavior are regulated by various mechanisms and are important in natural selection.
4.B.2: Cooperative interactions within organisms promote efficiency in the use of energy and matter.

Recall It

Recall that the vascular tissues xylem and phloem move water and minerals from the roots to the leaves of a plant. The ability for even the tallest trees depend on the properties of water and transpiration to draw water all the way to the top. Water potential is lowest at the leaves as water escapes through the stomata via evaporation, so water is pulled via cohesion and adhesion up the water column. This known as the cohesion-tension model of xylem transport. The opening and closing of stomata play an important role in water transport. Phloem tissue that transfers the products of photosynthesis is described by the pressure-flow model. Phloem can travel in any direction. As leaves make sugar, that sugar is co-transported with hydrogen ions down their concentration gradients to a sink- a place of storage.

Review It

Describe the relationship between source and sink.

Identify the mechanism base on its description:

Description	Mechanism
Sugar moves from source to sink	
Water molecules cling together and form a chain	
Water enters the root	
Water molecules stick to xylem	
Water moves from one place to another	
Water and minerals travel up xylem	

Use It

Explain the process occurring in the following diagram.

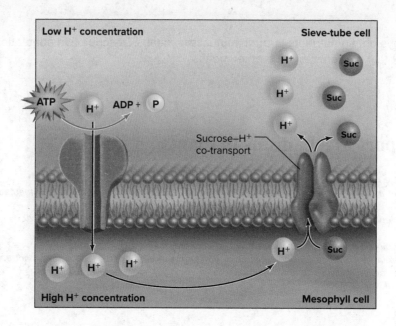

Determine if the following will make a stomata open (O) or close (C).

External carbon dioxide levels increase

Sunlight shines on the stomata

Potassium ions enter the guard cells

Guard cells take up water

Abscisic acid is produced by wilting leaves

Summarize It

Why do plants grow better when mycorrhizae are present?

Explain how water moves from the soil to the roots to the shoots and out of the leaves of a plant through changes in water potential and through transpiration.

Following *the* Big Ideas

AP Essential Knowledge	Chapter Section
BIG IDEA 2 **2.C.1** Organisms use feedback mechanisms to maintain their internal environments and respond to external environmental changes.	26.1
2.C.2 Organisms respond to changes in their external environment.	26.2, 26.3
2.D.4 Plants and animals have a variety of chemical defenses against infections that affect dynamic homeostasis.	26.1
2.E.2 Timing and coordination of physiological events are regulated by multiple mechanisms.	26.1, 26.2, 26.3
BIG IDEA 3 **3.B.2** A variety of intercellular and intracellular signal transmissions mediate gene expression.	26.1
3.D.1 Cell communication processes share common features that reflect a shared evolutionary history.	26.1
3.D.3 Signal transduction pathways link signal reception with cellular responses.	26.1

Chapter Overview

Plants respond to their environment, including other organisms, in a variety of ways. Environmental stimuli, including light, carbon dioxide, gravity, pathogens, drought, and physical touch can all elicit a response in a plant. Plants secrete hormones to target cells, changing patterns of growth. This allows them to respond to their environment. Ability for plants to sense change are found in many specialized cells and chemicals, including phytochromes and other the light sensing pigments.

26.1 Plant Hormones

Essential Knowledge covered
2.C.1: *Organisms use feedback mechanisms to maintain their internal environments and respond to external environmental change.*
2.D.4: *Plants and animals have a variety of chemical defenses against infections that affect dynamic homeostasis.*
2.E.2: *Timing and coordination of physiological events are regulated by multiple mechanisms.*
3.B.2: *A variety of intercellular and intracellular signal transmission mediate gene expression.*
3.D.1: *Cell communication processes share common features that reflect a shared evolutionary history.*
3.D.3: *Signal transduction pathway link signal reception with cellular response.*

Recall It

Plants contain many different types of hormones. These hormones are important chemical signals that coordinate cell behavior and response. This section introduced you to five classes of plant hormones: (1) auxins, (2) gibberellins, (3) cytokinins, (4) abscisic acid, and (5) ethylene. Hormones bring about change through signal transduction. A hormone is expressed under a certain environmental condition, then activates a series of proteins that promotes a cellular process such as plant growth, plant defense, or some sort of plant development.

Review It

Identify the plant hormone that triggers the response:

Response	Hormone
Increases the length of a stem	
Causes a plant to grow taller and taller	
Promotes cell division	
Closes stomata	

Use It

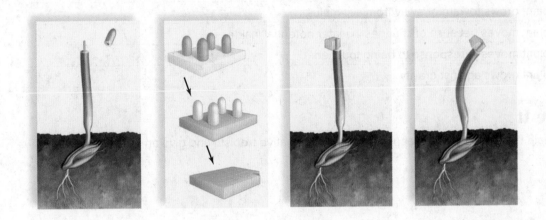

The figure above illustrates an important study that determined how phototrophism occurs in seedlings. Explain what is happening in the illustration and the conclusion of the study.

Two insects begin to feed on a plant leaf. The first insect takes a bite without any problem. The second insect takes a bite and experiences a terrible sensation. Why did the first insect have no problem eating the leaf but the second insect does?

26.2 Plant Growth and Movement Responses

Essential Knowledge covered
2.C.2: Organisms respond to changes in their external environment.
2.E.2: Timing and coordination of physiological events are regulated by multiple mechanisms.

Recall It

Growth to or from a unidirectional stimulus is called a tropism. There are many types of tropisms in plants: phototropism, thigmotropism, gravitropism. Turgor movement, on the other hand, is a change in plant cells caused by a change in turgor pressure. Turgor movements do not involve the growth of the plant.

Review It

Identify the plant movement based on the description.

Description	Movement
A plant grows towards light	
A plant grows toward or away from a stimulus	
A plant moves because of changes in water potential inside	
A plant moves in response to being touched	
A plant grows against gravity	

Use It

Explain the difference between a positive and negative tropism and give an example of each.

The sensitive plant, *Mimosa pudica*, will collapse its whole leaf within a second after you touch it. Explain how it is able to do this.

26.3 Plant Responses to Phytochrome

Essential Knowledge covered
2.C.2: Organisms respond to changes in their external environment.
2.E.2: Timing and coordination of physiological events are regulated by multiple mechanisms.

Recall It

A circadian rhythm refers to a cycle of activity in organisms during a 24-hour period. The mechanism that keep the circadian rhythm maintained in the absence of sunlight is an example of a biological clock. A physiological response that is dependent on the length of day or night is called photoperiodism. Short-day plants require longer period of dark than light in order to flower. Long-day plants require a shorter period of dark than light to flower. Plants that are not dependent on day length are day-neutral. The activity of the leaf pigment phytochrome is necessary for photoperiodism to occur. Plants grown in the dark tend to be etiolated, which means their stems are elongated and their leaves are small and yellow.

Review It

List two ways in which plants respond to changes in light.

Use It

A scientist is trying to classify a plant as a short-day or long-day plant. If the plant still blooms even when the night cycle is interrupted with a bright flash of light, what type of plant is it?

Suppose you pick up a packet of seeds. On the back, it says to only partly cover the seeds with soil when planting and then space the seedling when they germinate 10-12 inches apart. Explain the scientific reasoning behind these instructions.

The figure below shows the conversion of phytochrome from the inactive form to the active form. Circle the mechanism which initiates the change and explain two functions of the active phytochrome.

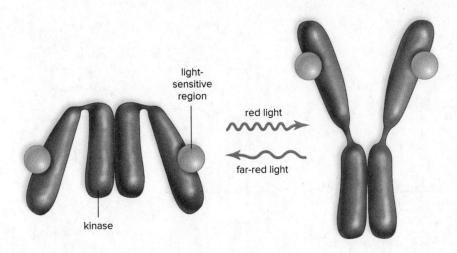

light-sensitive region

red light

far-red light

kinase

Summarize It

Imagine you are a grocery store owner, and you are shipped green tomatoes. How is it possible to ripen the tomatoes even though they are already off the vine?

How does phototropism occur in plants?

What is an example of a chemical deterrent plants have evolved to keep predators or competition away?

27 Flowering Plants: Reproduction

Following *the* Big Ideas

Chapter Overview

The flower of an angiosperm is more than just a pretty accessory – it is fundamental in the ability for the plant to reproduce sexually. The flower produces microspores and megaspores and protects the gametophyte. After successful pollination, an embryo develops in the ovule, which eventually matures into a seed. The ovary becomes a fruit. To enhance pollination rates and seed dispersal strategies, a staggering array of diversity in flowers and fruits have evolved in flowering plants and coevolved with other organisms. Many flowering plants can also reproduce through asexual reproduction, creating clones.

27.1 Sexual Reproductive Strategies
Extending Knowledge

Recall It

Variation in flowering plant offspring is produced through sexual reproduction. An alternation of generations is followed as a diploid sporophyte produces flowers, and the flowers produce haploid microspores and megaspores. The microspores and megaspores develop into male and female gametophytes – pollen grains and embryo sacs. When the pollen grain enters and fertilizes the female gametophyte, a diploid embryo develops. No external water is required to bring about fertilization in a flowering plant, as the gametophytes are well protected from desiccation. Pollination is the successful transfer of pollen into an embryo sac. Some plants self-pollinate, while others have coevolved with other organisms, such as bees, to facilitate the movement of pollen from one plant to another.

Review It

Identify the following structures in the diagram below:

Seed, pollen grain, microspore, ovule, megaspore, ovary, anther, sporophyte, embryo sac, seed

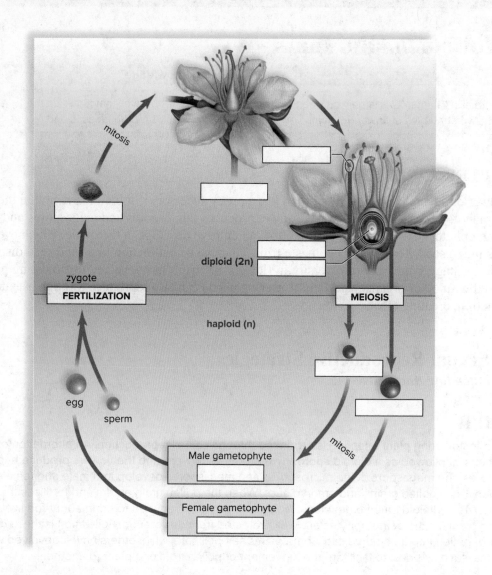

Describe the events that must occur before a pollen grain can become a mature male gametophyte.

Describe an example of coevolution between an angiosperm and a pollinator that you read about in this section.

27.2 Seed Development
Extending Knowledge

Recall It

After fertilization, a plant embryo develops in a series of stages; the zygote and proembryo stage, the globular stage, and the heart and torpedo stage. The final result is a seed with a protective seed coat, an area that contributes to shoot formation, an area that contributes to root formation, and a food supply called endosperm.

Review It

List and describe three parts of a seed.

What is the importance of the seed coat?

27.3 Fruit Types and Seed Dispersal
Extending Knowledge

Recall It

A fruit is a mature ovary that can also contain other flower parts. There are different kinds of fruit: a simple fruit, a compound fruit, an aggregate fruit, and a multiple fruit. These fruit types are defined by how many ovaries the fruit has and how the ovaries are arranged. Fruits are also classified by their texture, which is based on the mature pericarp. Fruits are dispersed in many different ways, including by air, by animal, and by ejection. Following dispersal, seeds germinate to form a seedling.

Review It

What is your favorite fruit? Classify your favorite fruit based on its composition.

Suppose you pick up a packet of seeds. On the back, it says to place the packet in the refrigerator for one week before planting in the ground. Explain the scientific reason behind these instructions.

27.4 Asexual Reproductive Strategies
Extending Knowledge

Recall It

Asexual reproduction results in clone offspring. Many plants are capable of asexual reproduction. Stolons are horizontal stems that produce clones. Rhizomes are underground stems that produce new plants asexually. Tissue culture is plant propagation done under sterile conditions. Plant cells are totipotent, meaning a plant cell can develop into an entire plant. As such, tissue culture in plants has become an important method to propagate fruits and vegetables for market and also in plant conservation. Chemicals are also extracted from plant cells by scientists for medicinal use.

Review It

Compare and contrast stolon and rhizomes.

Provide one advantage and one disadvantage for asexual reproduction in plants.

AP CHAPTER SUMMARY

Summarize It

How are the reproductive strategies of flowering plants adapted to a terrestrial lifestyle?

Your friend is growing plants that produce yellow, blue, and white flowers, which open during the day. Your friend claims they are bat-pollinated. Is your friend correct? What do you think pollinates these flowers?

UNIT FIVE: PLANT EVOLUTION AND BIOLOGY

Chapter 23 Plant Evolution and Diversity ● Chapter 24 Flowering Plants: Structure and Organization ● Chapter 25 Flowering Plants: Nutrition and Transport ● Chapter 26 Flowering Plants: Control of Growth Responses ● Chapter 27 Flowering Plants: Reproduction

Multiple Choice Questions

Directions: For each question or incomplete statement chose one of the four suggested answers or completions listed below.

1. Potato mosaic virus causes loss of pigmentation in plant cells. Once it enters the plant cell, it moves between plant cells using:

   (A) Stomata

   (B) Plasmodemata

   (C) Neurotransmitters

   (D) Chloroplasts

2. A potted plant placed on a window sill leans towards the light. What tropism is responsible for the movement of this plant?

   (A) Chemotropism

   (B) Thigmotropism

   (C) Phototropism

   (D) Gravitropism

3. Some waterlilies bloom at night, while other waterlilies bloom during the day. A recent study found that the flowers of night-blooming waterlilies produce aromatic ethers and ester volatile organic compounds (VOCs); whereas day-blooming species emit aromatic alcohol and ether VOCs. The difference in the types of VOCs is probably most likely linked to

   (A) VOC production in the light, versus VOC production in the dark.

   (B) Pollinator selection.

   (C) Convergent evolution.

   (D) The color of the waterlily flower.

4. Dahlias are an ornamental plant found in the backyards of many home gardeners. There are a staggering 20,000+ number of dahlia breeds. The most likely mechanism for this dazzling array of genetic diversity is most likely the result of:

(A) Polyploidy

(B) Genetic drift

(C) Adaptive radiation

(D) Mutation

5. Which of the following describes a pathogen-specific immune response in plants?

(A) Plants have cell walls that are often covered in waxy coatings.

(B) Plants have cell-surface receptors that can identify certain pathogen chracteristsics and trigger the production of toxic chemicals.

(C) Plants can produce toxic chemicals.

(D) Plants do not have pathogen-specific immune response systems.

6. Many tomatoes found on supermarket shelves do not ripen naturally. These tomatoes have been genetically altered to:

(A) Produce ethylene

(B) Not to produce ethylene

(C) Have red pigment even when unripe

(D) Produce abscisic acid (ABA)

Free Response Questions

Directions: Read the questions carefully and completely. Then, plan your answer and write your response in the space provide. Write your answer out in paragraph form.

1. Scientists studied the effect of heat shock and drought stress on tobacco plants in a laboratory setting. They measured both the stomatal conductance (A) and leaf temperature (B) in a series of experiments. Stomatal conductance is an estimation of the rate of gas exchange through the leaf stomata.

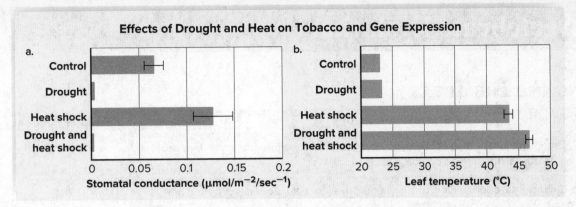

Effects of Drought and Heat on Tobacco and Gene Expression

Data obtained from: Rizhsky, Ludmila, Hongjian Liang, and Ron Mittler. 2002. The combined effect of drought stress and heat shock on gene expression in tobacco. Plant Physiology 130(3):1143–1151.

(a) Using your knowledge of plant biology and plant physiology, **provide a plausible hypothesis** as to why stomatal conductance would be lower under Drought conditions, and Drought + Heat Shock treatments, but not under heat shock treatment alone.

(b) Illustrate the process of what stomatal conductance may look like.

(c) Predict what the physiological effects of Drought + Heat Shock may be on tobacco plants.

2. Two plants of the same species were grown in a greenhouse setting in pots under identical conditions. In one pot (Pot A), mycorrhizae was added, but the soil in the other pot (Pot B) had been sterilized. After two months, the plants were measured. The biomass of Pot A was twice as great as Pot B. Explain why the plant in Pot B had less biomass.

FOLLOWING *the* BIG IDEAS

	AP Essential Knowledge	Chapter Section
BIG IDEA 1	**1.B.1** Organisms share many conserved core processes and features that evolved and are widely distributed among organisms today.	28.1
BIG IDEA 2	**2.E.1** Timing and coordination of physiological events are regulated by multiple mechanisms.	28.1
BIG IDEA 3	**3.B.2** A variety of intercellular and intracellular signal transmissions mediate gene expression.	28.1

CHAPTER OVERVIEW

All animals most likely evolved from a choanoflagellate-like protist. Invertebrates are animals which do not have backbones. The simplest invertebrates: sponges, comb jellies, and cnidarians are the only animals that are not triploblastic. Spiralla is a group of invertebrates that includes lophotrochozoans- animals such as mollusks, cephalopods, and annelids, and platyzoans, which encompass rotifers and flatworms. Arthropods and roundworms are invertebrates that periodically shed a cuticle, and echinoderms are the only invertebrates that share the deuterostome pattern of development with chordates.

28.1 Evolution of Animals

Essential Knowledge covered
1.B.1:   *Organisms share many conserved core processes and features that evolved and are widely distributed among organisms today.*
2.E.2:   *Timing and coordination of physiological events are regulated by multiple mechanisms.*

Recall It

Animals are multicellular eukaryotes that lack cell walls. Most have nerves and muscles, and reproduce sexually. Vertebrates have a backbone, while invertebrates do not. It is currently thought that animals descended from a protist ancestor. Slight shifts in the expression of homeotic genes may be responsible for the vast diversity we see in the animal kingdom today. Different forms of symmetry are seen in the animal kingdom: (1) asymmetry in sponges, (2) radial symmetry in cnidarians and ctenophores, and (3) bilateral symmetry in the rest. Animals are either protostomes or deuterostomes and are composed of either two or three germ layers.

28.1 Evolution of Animals *continued*

Review It

Define or identify the following terms that are associated with the evolution of animals:

Term	Definition
choanoflagellates	
invertebrates	
cephalization	
	The first developmental event after fertilization
	The anus develops prior to the mouth
radial symmetry	
protostomes	
	Defined right and left halves
coelom	
	Animals that have a spinal cord at some stage of their lives
asymmetry	
germ layers	
	A hollow sphere of cells
	A pattern of similarity observed in living organisms

Use It

Describe the differences and similarities between protostome and deuterostome development and give an example animal for each.

After the embryo of an animal with bilateral symmetry develops an anterior and posterior region, describe what genes determine the late of that animal's segmentation pattern.

28.2 The Simplest Invertebrates
Extending Knowledge

Recall It

Sponges, cnidarians, and ctenophores are considered the simplest invertebrates. Sponges are thought to be the oldest invertebrates. They lack true tissue and cellular organization. Ctenophora or comb jellies are mostly free-swimming marine invertebrates that exhibit greater organization than sponges. Cndiarians have a gastrovascular cavity that acts as a supportive hydrostatic skeleton. Cnidarians, such as the hydra, also have nerve nets. Ctenophores and cnidarians have true tissues, two germ layers, and are radially symmetrical as adults.

28.2 The Simplest Invertebrates *continued*
Extending Knowledge

Review It

Identify the structure as either belonging to a sponge (S), comb jelly (CJ) or cnidarian (C).

spicule

hydrostatic skeleton

mesoglea

gastrovascular cavity

nematocyst

Describe how a sponge obtains nutrients.

Hydras have something called a *nerve net*. Describe what it is and how it works.

28.3 Diversity Among the Lophotrochozoans
Extending Knowledge

Recall It

The Spiralians contains two groups: the Lophotrochozoans and the Platyzoa. Members of the group Spiralia have bilaterally symmetry at some point in development, spiral cleavage, and three germ layers. Adults have organs. The Lophotrochozoans contains molluscs, phoronids, tapeworms, brachiopoda, rotifers, flatworms, annelids, and bryozoans. The Platyzoans contain flatworms and rotifers.

Review It

The organisms of the lophotrochozoa are very diverse in their anatomy and life cycles. What are some characteristics they all share in common?

Describe how parasitic flatworms have evolved to live within their hosts.

28.4 Diversity of the Ecdysozoans
Extending Knowledge

Recall It

Roundworms and arthropods belong to the group Ecdyosoan. Ecdyosozoans have an outer covering called a cuticle, which is both a protective shell and for structural support. The phylum Nematoda contain roundworms, nonsegmented worms which have a body cavity called a pseudocoelom. Many organisms including insects, crustaceans, and spiders are arthropods. Arthropods have distinct features including an exoskeleton, nervous system, segmentation, and respiratory organs, such as air tubes called tracheae. Some arthropods undergo a drastic change during their life cycle called metamorphosis.

Review It

The organisms of the Ecdysozoans are very diverse in their anatomy and life cycles. What are some characteristics they all share in common?

What are some features which have led to the evolutionary success of arthropods?

28.5 Invertebrate Deuterostomes
Extending Knowledge

Recall It

Echinoderms are invertebrate deuterostomes which exhibit radial symmetry as adults but bilateral symmetry in the larval stage. Echinoderms have unique water vascular systems that aid in locomotion, gas exchange, and feeding. Echinoderms do not have a respiratory, excretory, or circulatory system.

Review It

List two ways sea stars and chordates are related.

A sea star is a member of the Phylum Echinodermata. Write a short paragraph describing its life cycle, morphology, and physiology.

Summarize It

What features are common to all animals, regardless of their complexity, lifestyle, and habitat?

How do *HOX* genes orchestrate the development of the body plan of animals, while also providing evidence for a common ancestor?

FOLLOWING *the* BIG IDEAS

AP Essential Knowledge	Chapter Section
BIG IDEA 1 **1.A.2** Natural selection acts on phenotypic variations in populations.	29.2, 29.4
1.B.1 Organisms share many conserved core processes and features that evolved and are widely distributed among organisms today.	29.6
1.B.2 Phylogenetic trees and cladograms are graphical representations (models) of evolutionary history that can be tested.	29.1, 29.5
BIG IDEA 2 **2.B.2** Growth and dynamic homeostasis are maintained by the constant movement of molecules across membranes.	29.3
2.C.2 Organisms respond to changes in their external environment.	29.3
2.D.2 Homeostatic mechanisms reflect both common ancestry and divergence due to adaptation in different environments.	29.3

CHAPTER OVERVIEW

Vertebrates are chordates with a backbone. All chordates have a notochord, a dorsal tubular nerve cord, pharyngeal gill pouches, and a postanal tail at some point during development. Vertebrates include fish, amphibians, reptiles, birds, and mammals. Fish, amphibians, and reptiles are ectotherms, while birds and mammals are endotherms. The evolutionary relationships between and within the chordates can be viewed through phylogeny based on shared and conserved morphological traits.

29.1 The Chordates

Essential Knowledge covered
1.B.1: Organisms share many conserved core processes and features that evolved and are widely distributed among organisms today.
1.B.2: Phylogenetic trees and cladograms are graphical representations (models) of evolutionary history that can be tested.

Recall It

At some point in development, all chordates possess the following four traits: (1) a notochord, (2) a dorsal tubular nerve cord, (3) pharyngeal gill pouches, and (4) a postanal tail. These four traits link all chordates back to a single common ancestor. Most chordates also have a backbone, but some chordates are nonvertebrate chordates. These include the cephalochordates (lancelets) and urochordates (sea squirts). Sea squirts are thought to be directly related to the vertebrates.

Review It

What do cephalochordates and urochorates have in common?

If you were to draw a phylogenic tree of the chordates, what trait would the tunicates and lancelets share with the vertebrates, and which trait would they lack?

29.1 The Chordates *continued*

Use It

Identify and describe the structures numbered in the diagram below.

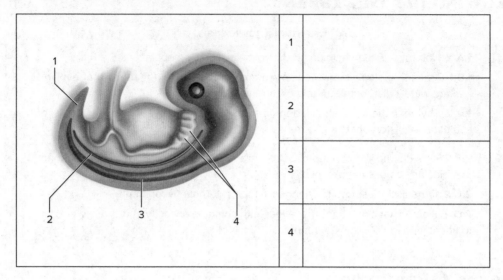

1	
2	
3	
4	

The structures in the diagram above are all seen at some point during development of all chordates. What does this indicate about all chordates?

29.2 The Vertebrates

Essential Knowledge covered
1.A.4: Biological evolution is supported by scientific evidence from many disciplines, including mathematics.
4.A.4: Organisms exhibit complex properties due to interactions between their constituent parts.

Recall It

Vertebrates are chordates with four distinct characteristics: (1) a vertebral column, (2) a skull, (3) an endoskeleton, and (4) internal organs. Ample evidence in the fossil record shows us that jawless and jawed fishes were the first vertebrates to appear, followed by amphibians. Reptiles produce an amniotic egg, as do mammals, making these vertebrates the only chordates fully adapted to land.

Review It

How are vertebrates different from sea squirts or lancelets?

29.2 The Vertebrates *continued*

Provide the definitions for the following terms:

Term	Definition
Tetrapod	
Gnathostome	
Amniote	
Cephalization	

Use It

Describe some traits found in fish fossils from the Ordovician which showed characteristics of adaptation to land.

List three benefits of having an endoskeleton.

29.3 The Fishes

Essential Knowledge covered
2.B.2: Growth and dynamic homeostasis are maintained by the constant movement of molecules across membranes.
2.C.2: Organisms respond to changes in their external environment.
2.D.2: Homeostatic mechanisms reflect both common ancestry and divergence due to adaptation in different environments.

Recall It

The largest group of vertebrates are the fishes. Jawless fishes evolved first, with the jaw evolving later from the gill arch. Fishes are ectotherms, meaning they depend on the environment to regulate their temperature. Jawed fishes are classified as either cartilaginous or bony, based on the makeup of their endoskeleton. Lobe-finned fishes are thought to be most closely related to the amphibian.

Review It

Identify the fishes based on the description.

Description	Fishes
sharks, rays, skates: fishes with cartilaginous skeletons and lack gill covers	
the majority of living vertebrates	
extinct jawed fishes for the Devonian, thought to be the early ancestors of sharks and bony fishes	
bony fishes with fleshy fins supported by bones; includes lungfish	
bony fishes with fan-shaped fins	
agnathans	

29.3 The Fishes *continued*

Use It

How do most fishes regulate their temperature?

What is a swim bladder and how does it work?

During field work, you come across a species of fish you've never seen before. How would you classify the fish as cartilaginous, bony, or lobe-finned?

29.4 The Amphibians

Essential Knowledge covered
1.A.2: Natural selection acts on phenotypic variations in populations.

Recall It

Amphibians are unique in that they are both able to live on land and in water. They have several characteristics which set them apart from fishes and reptiles. Amphibians are typically tetrapods, have smooth, moist skin, small lungs, and sense organs developed for use on land as opposed to in water. Like fish, amphibians are ectotherms. Amphibians have a single-loop circulatory system. The majority of amphibians return to water to reproduce, where they hatch aquatic larvae that undergo a metamorphosis and eventually return to land.

Review It

Identify the name of the characteristic that defines and amphibian.

Description	Characteristic
plays an active role in water balance, respiration, and thermal regulation	
eggs and sperm are deposited in the water	
sight, hearing, smell	
three chambered heart with a single ventricle and two atria	
depend on environment to regulate internal temperature	
developed pelvic and pectoral girdle	
respiratory organs which are supplemented with cutaneous respiration	

29.4 The Amphibians *continued*

Use It

Amphibians have become very diverse. Describe an adaptation that has allowed a particular amphibian to thrive in its environment.

29.5 The Reptiles

Essential Knowledge covered
1.B.2: Phylogenetic trees and cladograms are graphical representations (models) of evolutionary history that can be tested.

Recall It

Reptiles are fully adapted to land. They have paired limbs, thick skin, efficient breathing, circulation, and excretion. Most reptiles are ectotherms. Fertilization occurs internally, preventing any desiccation of sperm or egg. Currently, reptiles also now include birds, as more evolutionary relationships are elucidated. Birds are unlike other reptiles in that they are endotherms. Birds have also become well adapted to flight, with feathers, modified respiration, and a modified skeleton.

Review It

How do reptiles maintain their body temperature? What is the exception?

Use It

Turtles, birds, snakes, and crocodiles are all reptiles. Using the information presented in this section, draw a phylogenic tree showing how they could possibly be related.

29.6 The Mammals

Essential Knowledge covered
1.B.1: Organisms share many conserved core processes and features that evolved and are widely distributed among organisms today.
4.B.4: Distribution of local and global ecosystems change over time.

29.6 The Mammals *continued*

Recall It

The mammals are animals with hair and mammary glands. Most animals produce young that are born alive after a period of internal development. Mammals are broken into three lineages: (1) the monotremes, (2) the marsupials, and (3) the placentals. Monotremes are unique in that they do not have nipples and do not have internal embryonic development; they lay eggs. Marsupials do bare live offspring, but newborns are immature and spend some time inside a pouch on their mother's abdomen continuing to develop. The placentals, as per their name, have a specialize organ called a placenta, which gives all nutrients a developing offspring needs while in utero to be born in a more developed state than marsupial offspring.

Review It

Describe the three characteristics that are unique to mammals.

List the three mammalian lineages.

Define *placenta*.

Use It

Compare and contrast monotremes and marsupials.

Placental mammals are the most dominant group of mammals. What event allowed for this group to become so diverse?

AP CHAPTER SUMMARY

Summarize It

What characteristics distinguish vertebrates from other animals?

How does vertebrate evolution relate to the Earth's changing environments?

30 Evolution of Primates

FOLLOWING *the* BIG IDEAS

AP Essential Knowledge	Chapter Section
BIG IDEA 1 **1.A.4** Biological evolution is supported by scientific evidence from many disciplines, including mathematics	30.1
1.B.1 Organisms share many conserved core processes and features that evolved and are widely distributed among organisms today.	30.1
1.B.2 Phylogenetic trees and cladograms are graphical representations (models) of evolutionary history that can be tested.	30.1, 30.3
1.C.1 Speciation and extinction have occurred throughout the Earth's history.	30.3
1.C.2 Speciation may occur when two populations become reproductively isolated from each other.	30.3

CHAPTER OVERVIEW

Primates include prosimians, monkeys, apes, and humans. Most primates are arboreal with an evolutionary trend to bipedalism. Fossil and molecular data have provided many interesting relationships within the primates. Humans are most closely related to chimpanzees. The human lineage is well described through the fossil record, as well. Several groups of hominins preceded the evolution of the modern human.

30.1 Evolution of Primates

Essential Knowledge covered
1.A.4: Biological evolution is supported by scientific evidence from many disciplines, including mathematics.
1.B.1: Organisms share many conserved core processes and features that evolved and are widely distributed among organisms today.
1.B.2: Phylogenetic trees and cladograms are graphical representations (models) of evolutionary history that can be tested.

Recall It

Primates include prosimians, monkeys, apes, and humans. They have several traits which have made them well adapted to an arboreal life: mobile forelimbs, mobile hindlimbs, and stereoscopic vision. Many primates also have opposable thumbs and toes so that they are able to grasp things precisely. Primates have large, complex brains, and a general reduction in the rate of reproduction. Primates are further classified into Prosimians, Anthropoids, Hominoids, Hominids, and Hominins.

Review It

Describe how primates are adapted for an arboreal life.

Use a phylogenic tree to describe how chimpanzees, tarsiers, humans, gorillas, capuchin monkeys, rhesus monkeys, and lemurs are related.

Use It

Proconsul is a fossil from about 35 MYA. What does Proconsul represent and how does it compare to a current monkey skeleton?

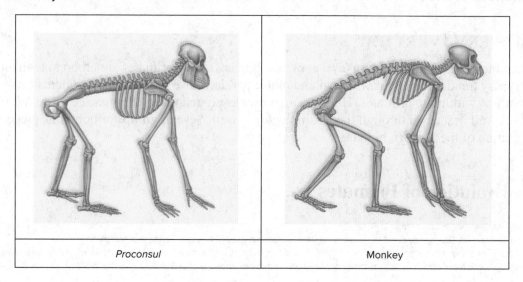

| Proconsul | Monkey |

30.2 Evolution of Humanlike Hominins
Extending Knowledge

Recall It

Molecular and fossil data suggest that hominin evolution began about 5 MYA. Bipedalism was a major change in the hominin lineage. Bipedalism is advantageous for individuals carrying offspring and foraging. As hominin evolved, they began to stand more upright and developed larger brains, although this did not occur all at once, but was rather a product of mosaic evolution. Mosaic evolution occurs when different body parts change at different rates and different times.

30.2 Evolution of Humanlike Hominins *continued*
Extending Knowledge

Review It

Identify the hominin.

Description	Hominin
• Slightly larger than a chimpanzee • Small head, brain ~370 to 515 cc • Walked erect, may have eaten meat • Used stone tools	
• About the size of a chimpanzee • Small head, brain ~300 to 350 cc • Fossil remnants date back to 5.6 MYA • Walked erect but spent most of the time in trees • Had opposable big toes	

How are *Homo sapiens* different from Ardipithecines and Australopithecines? What traits do they have in common?

30.3 Evolution of Early Genus *Homo*

Essential Knowledge covered
1.B.2: Phylogenetic trees and cladograms are graphical representations (models) of evolutionary history that can be tested.
1.C.1: Speciation and extinction have occurred throughout the Earth's history.
1.C.2: Speciation may occur when two populations become reproductively isolated from each other.

Recall It

Several early *Homo* species appear in the fossil record. These species include: *Homo habilis*, *Homo rudolfensis*, *Homo ergaster*, *Homo erectus*, and *Homo floresiensis*. These species appear in the fossil record during different times and have distinct morphological traits.

Review It

Use the information given to you in this section to describe the traits and most probable time period of the early *Homo* species.

Homo species	Traits	Time Period
Homo floresiensis		
Homo ergaster		
Homo habilis and *Homo rudolfensis*		

30.3 Evolution of Early Genus *Homo* continued

Use It

Describe the origins and speciation of *Homo floresiensis*.

30.4 Evolution of Later Genus *Homo*
Extending Knowledge

Recall It

The replacement model or out-of-Africa-hypothesis is the most widely accepted hypothesis for the evolution of modern humans. This hypothesis proposes the modern human evolved from earlier *Homo* species found only from Africa, before migrating to Asia and Europe and displacing the other *Homo* species found there. This would have included the species *Homo neandertalensis*, whose fossils have been found throughout Europe. Although, interbreeding may have occurred between *Homo neandertalensis* and *Homo sapiens* (modern humans). Cro-Magons are the oldest *Homo sapiens*. Modern *Homo sapiens* are quite diverse externally as different populations evolved as adaption to local conditions but genetically are very similar.

Review It

Fill in the most probable word or description associated with the later genus *Homo*.

Term	Description
biocultural evolution	
	Oldest fossils of *Homo sapiens*
Denisovans	
hunter-gatherers	
	archaic humans that lived 200,000-300,000 years ago

Modern humans have a wide variation in phenotype. What are two hypotheses that explain these variations?

While populations of humans across the world do show variation in phenotypes, what evidence is there that we all arose from a common ancestor?

Summarize It

What can the fossil record and comparative genomics tell us about human evolution?

UNIT SIX: VERTEBRATE EVOLUTION

Chapter 28 Invertebrate Evolution • Chapter 29 Vertebrate Evolution • Chapter 30 Human Evolution

Multiple Choice Questions

Directions: For each question or incomplete statement chose one of the four suggested answers or completions listed below.

Use the following phylogenetic tree to answer questions 1 and 2

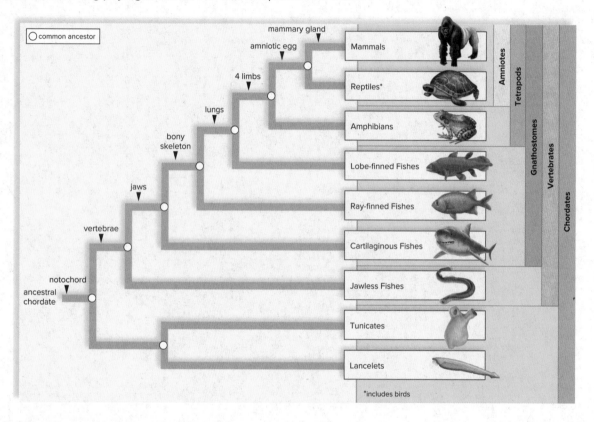

*includes birds

1. According to the figure, which trait is distinctive to all Chordates?

    (A) Vertebrae

    (B) Notochords

    (C) Mammary Glands

    (D) Lungs

2. Amphibians most likely evolved from

    (A) Lobe-finned fishes

    (B) Ray-finned fishes

    (C) Jawless fishes

    (D) Reptiles

3. Humans have 46 chromosomes, and chimpanzees have 48, but our genomes differ by only 1.5%. What hypothesis best explains this phenomenon?

    (A) The two extra chromosomes that chimpanzees have contain all of the different genes.

    (B) Similar genetics between humans and chimpanzees is the result of convergent evolution.

    (C) Human chromosome 2 is actually a fusion of two chimpanzee chromosomes.

    (D) Chimpanzees developed two extra chromosomes through nondisjunction as they evolved from humans.

4. How has the evolution of invertebrate genomes led to the amazing diversity of body forms observed?

    (A) Genomes have gotten more complex in more complex invertebrates.

    (B) Genomes increase in size in more complex invertebrates.

    (C) Genomes do not have any impact on invertebrate body form diversity.

    (D) Mutations in the *HOX* genes have led to diversity in body plans in invertebrates.

5. In which way are insects similar to plants?

    (A) Insects are heterotrophs.

    (B) Insects are eukaryotes.

    (C) Insects are animals.

    (D) Insects lack cell walls.

Free Response Questions

Directions: Read the questions carefully and completely. Then, plan your answer and write your response in the space provide. Write your answer out in paragraph form.

1. In a short essay, **describe** the four characteristics of chordates. **Explain** how these traits support the common shared ancestry of nonvertebrate and vertebrate chordates.

2. Using the following chart, **create a cladogram** showing the relationships between these mammals.

	Milk	Nipples	Birth	Uterus	Placenta	Birth Canal
Monotremes	Yes	No	Eggs	None	None	Cloaca
Marsupials	Yes	Yes	Live	True (>1)	Simple (yolk)	Vagina (>1)
Placentals	Yes	Yes	Live	True (1)	Complex (tissue)	Vagina (1)

(b) Whales, shrew, monkeys, and bats are both placental mammals. Describe what type of speciation evet must have occurred to give rise to so many different types of species?

FOLLOWING *the* BIG IDEAS

AP Essential Knowledge	Chapter Section
BIG IDEA 2 **2.B.2** Growth and dynamic homeostasis are maintained by the constant movement of molecules across membranes.	31.2
2.C.1 Organisms use feedback mechanisms to maintain their internal environments and respond to external environmental changes.	31.2, 31.3, 31.4
2.C.2 Organisms respond to changes in their external environments.	31.3, 31.4
2.D.1 All biological systems from cells and organisms to populations, communities and ecosystems are affected by complex biotic and abiotic interactions involving exchange of matter and free energy.	31.2
2.D.2 Homeostatic mechanisms reflect both common ancestry and divergence due to adaptation in different environments.	31.1, 31.4
2.D.4 Plants and animals have a variety of chemical defenses against infections that affect dynamic homeostasis.	31.1
BIG IDEA 4 **4.A.4** Organisms exhibit complex properties due to interactions between their constituent parts.	31.4

CHAPTER OVERVIEW

Animals are highly organized organisms. Specialized cells are organized into tissues. Tissues include: (1) epithelial tissue, (2) connective tissue, (3) muscular tissue, and (4) nervous tissue. These tissues then are organized into organs. Organs work together to perform a particular function. All organs work together to perform homeostasis; the ability to maintain a relatively constant internal environment. Frequently, negative feedback systems are used to maintain homeostasis; however, examples of positive feedback are also found in organisms as well.

31.1 Types of Tissues

Essential Knowledge covered
2.D.2: Homeostatic mechanisms reflect both common ancestry and divergence due to adaptation in different environments.
2.D.4: Plants and animals have a variety of chemical defenses against infections that affect dynamic homeostasis.

Recall It

There are four major types of tissues found in the most complex animals. Epithelial tissues cover body surfaces, body cavities, and forms glands. Connective tissue binds and supports body parts. Muscular tissue moves the body and its parts. Nervous tissue receives stimuli and transmits nerve impulse. Each type of tissue has further classification based on structure, cell type, and function.

Review It

Define *tissue*.

31.1 Types of Tissues *continued*

Name the tissue type given its function.

Function	Tissue
binds and supports body parts	
receives stimuli and transmits nerve impulses	
covers body surfaces, lines body cavities and forms glands	
moves the body and its parts	

Use It

How does connective tissue fight off infections in the human body?

31.2 Organs, Organ Systems, and Body Cavities

Essential Knowledge covered
2.B.2: Growth and dynamic homeostasis are maintained by the constant movement of molecules across membranes.
2.D.1: All biological systems from cells and organisms to populations, communities and ecosystems are affected by complex biotic and abiotic interactions involving exchange of matter and free energy.

Recall It

Organs are made up of one or more tissue type. Individual organs play a role in a larger organ system. Organ systems play out larger roles that are fundamental in life processes. Each organs system in located in a particular region of the body. Vertebrates have two body cavities: (1) the dorsal cavity and (2) the ventral cavity. These cavities are further divided in the human body to contain specific organ systems.

Review It

The diagram below illustrates three body cavities found in the ventral cavity of the human. Identify the cavity and list the organs and organ systems found within these locations.

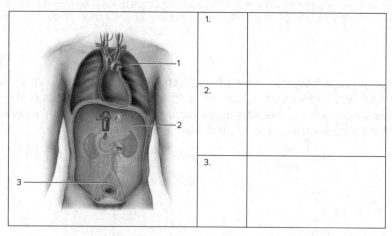

1.	
2.	
3.	

31.2 Organs, Organ Systems, and Body Cavities *continued*

Use It

Describe the relationships between an organ and an organ system.

List four organ systems and their corresponding life processes.

31.3 The Integumentary System

Essential Knowledge covered
2.C.1: *Organisms use feedback mechanisms to maintain their internal environments and respond to external environmental change.*
2.C.2: *Organisms respond to changes in their external environment.*

Recall It

The skin covers the human body and has two main regions. The epidermis is made up of stratified squamous epithelium, and the dermis is a region of dense fibrous connective tissue. In the epidermis, specialized cells called melanocytes produce a pigment called melanin; whereas, hair follicles, glands, and nails are largely part of the dermis. Humans have oil glands and sweat glands which secrete sebum and sweat, respectively.

Review It

List three functions of skin.

Compare and contrast an oil gland and a sweat gland.

Use It

Describe how sweating cools your body.

31.3 The Integumentary System *continued*

How does UV radiation from the sun effect melanin production in the skin?

31.4 Homeostasis

Essential Knowledge covered
2.C.1: Organisms use feedback mechanisms to maintain their internal environments and respond to external environmental change.
2.C.2: Organisms respond to changes in their external environment.
2.D.2: Homeostatic mechanisms reflect both common ancestry and divergence due to adaptation in different environments.

Recall It

Homeostatic regulation is the ability for organisms to maintain a relatively constant internal environment. Negative feedback is the primary homeostatic mechanism that maintains a constant value of a variable. When a sensor detects a change, a particular control center initiates an action and brings that environment back to its set point. Positive regulation also is found within organisms to a lesser degree, in which a control center brings about a continually greater change.

Review It

Identify the type of temperature homeostatic regulation.

Regulation	Definition
	Body temperature is regulated by a variety of internal mechanisms.
	Body temperature is regulated by the temperature of the external environment.

Given the organ system, describe a role it plays in homeostasis.

Organ	Homeostatic regulation
liver	
respiratory system	
kidneys	
digestive system	

Use It

Using the example of the regulation of temperature in the human body, describe what happens when the body falls below normal temperature. What type of feedback mechanism is this?

Summarize It

How do specialized tissues, organs, and organ systems allow animals to better adapt to their environment?

What is the difference between negative and positive feedback mechanisms in the regulation of homeostasis? Provide an example of each.

FOLLOWING *the* BIG IDEAS

CHAPTER OVERVIEW

Oxygen, nutrients, and waste products are moved around animals through some sort of circulatory system. As animals adapted to a terrestrial lifecycle, a closed circulatory system with blood evolved. The circulatory system is a critical link in all other organ systems.

32.1 Transport in Invertebrates
Extending Knowledge

Recall It

Most invertebrates have a circulatory system. Circulatory systems transport oxygen, nutrients, and waste products to and from cells. Blood and hemolymph are two types of circulatory fluids found in the circulatory system. Open circulatory systems contain hemolymph. Closed circulatory systems contain blood. Some invertebrates such as sponges, hydras, and flatworms lack a circulatory system because their thin body walls make gas and nutrient exchange possible without a separate system.

Review It

List two functions of a circulatory system.

What is the difference between an open and a closed circulatory system?

32.2 Transport in Vertebrates
Extending Knowledge

Recall It

All vertebrates have a closed circulatory system, known as a cardiovascular system. Blood is transported through the body in blood vessels. Fishes have one-circuit cardiovascular systems, while amphibians and most reptiles have a two-circuit system. In birds and mammals, a two-circuit system is also seen. Birds and mammals also have four-chambered hearts, as opposed to two chambers. Two-circuit systems are advantageous to land-dwellers as the heart pumps blood through a systemic circuit, as well as pumping blood through a pulmonary circuit, providing an efficient exchange of oxygen-rich and oxygen-poor blood.

Review It

Provide the correct term associated with circulation in vertebrates:

Definition	Term
Drain blood from the capillaries and join to form a vein	
Small arteries whose diameters are regulated by the nervous and endocrine system	
The closed circulatory system of vertebrates	
The two circuits of the vertebrate cardiovascular system	
Carry blood away from the heart	
Return blood to the heart	
Blood vessels that exchange materials with interstitial fluid	

Compare and contrast the circulatory pathway of a fish and a bird.

Describe the movement of blood and oxygen through the circulatory pathway in the space next to the illustration below.

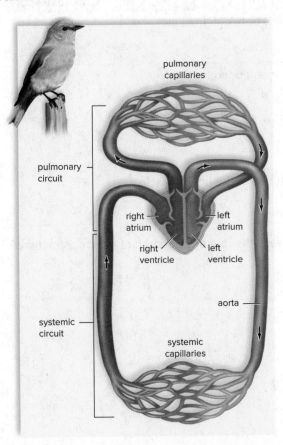

32.3 The Human Cardiovascular System
Extending Knowledge

Recall It

The structure and function of the human heart separates blood that is low in oxygen from blood that is high in oxygen. It supplies the body with oxygen-rich blood and sends blood that is low in oxygen and high in carbon dioxide to the lungs. The pumping of the heart keeps blood moving in the arteries, and skeletal muscle contractions against the veins forces blood through the veins. The cardiac cycle consists of an alternating systole and diastole of the atria and ventricles of the cardiovascular system. The volume of blood that the left ventricle pumps out per minute is known as the cardiac output. Blood pressure is measured as the amount of pressure it takes to stop the flow of blood through an artery. Many cardiovascular diseases in the United States are the result of hypertension—high blood pressure.

Review It

Fill in the missing phases of the cardiac cycle.

Time	Atria	Ventricles
0.15 sec		Diastole
0.30 sec	Diastole	
0.40 sec		Diastole

Identify the cardiovascular disease based on its description.

Description	Cardiovascular Disease
accumulation of soft masses of fatty materials between the inner linings of arteries which can trigger clots	
the blocking of an artery which causes a burning or squeezing sensation in the chest	
high blood pressure, narrowing of arteries	
the complete blocking of an artery in which part of the heart muscle dies due to lack of oxygen	

What internal systems can increase or decrease the rate and strength of heart contractions?

32.4 Blood
Extending Knowledge

Recall It

Blood consists of two main portions: (1) plasma and (2) the formed elements, such as cells and platelets. There are several plasma proteins, including albumin, which are the most plentiful, and antibodies, which are produced in the immune system. Red blood cells transport oxygen and help transport carbon dioxide. Blood types are defined by the type of antigens found on red blood cells and in the plasma. White blood cells help fight infection, and platelets aid in clotting.

Review It

What are T cell and B cells?

Draw a picture of how platelets and fibrin threads plug a punctured blood vessel.

Describe how horseshoe crab blood clotting can identify a bacterial contamination.

Compare and contrast the Rh and ABO system.

Summarize It

How did the evolution of the four-chambered heart allow animals to adapt to challenges of terrestrial living?

Describe the role of blood in the human body.

33 The Lymphatic and Immune Systems

FOLLOWING *the* BIG IDEAS

AP Essential Knowledge		Chapter Section
BIG IDEA 2	**2.D.3** Biological systems are affected by disruptions to their dynamic homeostasis.	33.3, 33.4, 33.5
	2.D.4 Plants and animals have a variety of chemical defenses against infections that affect dynamic homeostasis.	33.1, 33.2, 33.3, 33.4
	2.E.1 Timing and coordination of specific events are necessary for the normal development of an organisms, and these events are regulated by a variety of mechanisms.	33.4
BIG IDEA 3	**3.B.2** A variety of intercellular and intracellular signal transmissions mediate gene expression.	33.4
	3.D.2 Cells communicate with each other through direct contact with other cells or from a distance via chemical signaling.	33.2, 33.3, 33.4
	3.D.4 Changes in signal transduction pathways can alter cellular response.	33.4, 33.5
BIG IDEA 4	**4.C.1** Variation in molecular units provides cells with a wider range of functions.	33.4, 33.5

CHAPTER OVERVIEW

The immune system is a complicated system which protects us from pathogens and toxins. There are two forms of immunity covered in this chapter: (1) innate immunity and (2) adaptive immunity. Innate immunity is much older than adaptive immunity and can be found in even the "simplest" multicellular organisms. Adaptive immunity involves B cells and T cells, which have antigen receptors. These antigen receptors elicit a response when binding to a specific antigen.

33.1 Evolution of Immune Systems

Essential Knowledge covered
2.D.4: Plants and animals have a variety of chemical defenses against infections that affect dynamic homeostasis.

Recall It

Even simple multicellular organisms, such as cellular slime molds, exhibit signs of an immune system. Cells which develop protective functions against invaders but show no signs of increased response on repeat exposure are known as innate immunity. Most multicellular organisms have innate immunity. Others, such as vertebrate animals, also have adaptive immunity. This means there is an immunological memory of the initial invader, and an increased response to specific repeat invaders.

Review It

State the function of the immune system.

Define *antigen*.

33.1 Evolution of Immune Systems *continued*

Use It

Identify two differences between innate immunity and adaptive immunity.

Compare and contrast the immune system of a fruit fly and a human.

33.2 The Lymphatic System

Essential Knowledge covered
2.D.4: Plants and animals have a variety of chemical defenses against infections that affect dynamic homeostasis.
3.D.2: Cells communicate with each other through direct contact with other cells or from a distance via chemical signaling.

Recall It

The lymphatic system contributes to homeostasis in three ways. The lymphatic system absorbs excess interstitial fluid and absorbs fats, but it also produces lymphocytes. Lymphocytes include B and T cells. These cells respond to signs of infection or inflammation, and destroy foreign invaders. Lymphocytes develop in primary lymphoid organs, such as bone marrow and thymus. Secondary lymphoid organs, such as lymph nodes, are a site to which mature lymphocytes migrate. The spleen is another secondary lymphoid organ where macrophages remove old and defective blood cells.

Review It

Identify the component of the lymphatic system:

Definition	Component
a secondary lymphoid organ where lymph passes through and phagocytes engulf foreign debris and pathogens	
spongy, semisolid tissue where red blood cells are produced	
a secondary lymphoid organ where macrophages remove old and defective red blood cells	
tiny, closed-ended vessels found throughout the body which take up excess interstitial fluid	
cells which mature in the thymus and fight infection	
the one-way system that drains fluid from tissue and returns it to the cardiovascular system	
a primary lymphoid organ where T cells mature	
interstitial fluid	
lymphocytes which mature in the red bone marrow	

33.2 The Lymphatic System *continued*

Use It

How do mature lymphocytes fight infections in the human body?

Why is a person without a spleen more susceptible to certain types of infections?

33.3 Innate Immune Defenses

Essential Knowledge covered
2.D.3: Biological systems are affected by disruptions to their dynamic homeostasis.
2.D.4: Plants and animals have a variety of chemical defenses against infections that affect dynamic homeostasis.
3.D.2: Cells communicate with each other through direct contact with other cells or from a distance via chemical signaling.

Recall It

Innate immunity occurs without initiation and is not amplified by exposure. Innate immunity defenses include: physical and chemical barriers, inflammatory responses, phagocytes and natural killer cells, and protective proteins. While innate immune responses are immediate, there is no immunological memory of the attack. There is no recognition of repeat intruders and, therefore, no change in immune response.

Review It

Identify the component (or components) of the innate immune system:

Description	Component
phagocytes which leave the bloodstream and kill bacteria in tissue	
cytokines which affect the behavior of other cells	
large, granular lymphocytes that kill virus-infected or cancer cells on contact	
a series of events that occur after infection by a pathogen	
two long-lived types of phagocytic white blood cells	
chemical mediators released by damaged cells which cause capillaries to dilate and become more permeable	
blood plasma proteins present in the blood plasma that are activated by pathogens and aid in immune response	
phagocytes which can also launch attacks against animal parasites such as tapeworms	
cells which release chemical mediators such as histamine	

33.3 Innate Immune Defenses *continued*

Identify the four types of innate defenses pictured below.

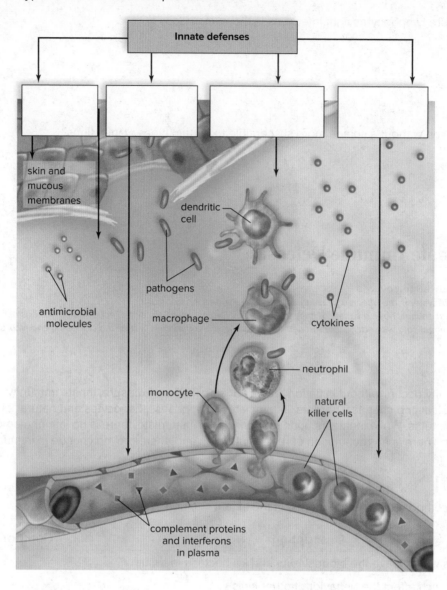

Use It

Imagine you drop something heavy on your big toe. Describe the response that occurs in the tissue.

What might happen to a cell without a MHC-1 molecule on its surface?

33.4 Adaptive Immune Defenses

Essential Knowledge covered
2.D.3: Biological systems are affected by disruptions to their dynamic homeostasis.
2.D.4: Plants and animals have a variety of chemical defenses against infections that affect dynamic homeostasis.
2.E.1: Timing and coordination of specific events are necessary for the normal development of an organism, and these events are regulated by a variety of mechanisms.
3.B.2: A variety of intercellular and intracellular signal transmission mediate gene expression.
3.D.2: Cells communicate with each other through direct contact with other cells or from a distance via chemical signaling.
3.D.4: Changes in signal transduction pathways can alter cellular response.
4.C.1: Variation in molecular units provides cells with a wide range of functions.

Recall It

Adaptive immunity takes a little bit longer to respond to infection than innate immunity, but once activated may last for years. B and T cells have antigen receptors which bind to antigens and trigger a response, such as the production of antibodies or cytotoxic T cells. Memory B or T cells are able to immediately recognize the antigen again in the future. Adaptive immunity can also be induced actively or passively. Immunizations contain antigens to which the immune system responds.

Review It

Provide the correct definition for each term associated with the adaptive immune system:

Term	Definition
immunoglobulins	
memory B cells	
helper T cells	
active immunity	
monoclonal antibodies	
immunization	

Define, in your own words, the clonal selection theory.

Use It

Draw and label a picture of a basic antibody. Describe how its antigen-binding sites plays a role in its function.

How are helper T cell and cytotoxic T cells different?

How does a T cell destroy a cell that has been infected with a virus?

33.5 Immune System Disorders and Adverse Reactions

Essential Knowledge covered
2.D.3: Biological systems are affected by disruptions to their dynamic homeostasis.
3.D.4: Changes in signal transduction pathways can alter cellular response.
4.C.1: Variation in molecular units provides cells with a wide range of functions.

Recall It

Sometimes, the immune system works against our best interest by becoming hypersensitive to allergens. This is known as an allergy. Anaphylactic shock is an immediate response to an allergen and is the most severe reaction to an allergen. Autoimmune disorders occur when the immune system mistakenly starts attacking its own body's cells or molecules. Other immune system disorders include immunodeficiencies; the lack of functioning T or B cells in an individual.

Review It

Identify the immune disorder or adverse reaction:

Description	Disorder
hypersensitivity to substances such as pollen or food that would not ordinarily cause a reaction	
a sudden and life-threating drop in blood pressure due to an immediate allergic response	
a reaction initiated by memory T cells at the site of allergen contact	
an immediate reaction to a substance such as pollen or food caused by IgE antibodies	
an inflammatory response that occurs in the lungs and causes wheezing	
when the immune system mistakenly attacks the body's own cells	

Use It

Describe how an allergic reaction occurs.

Choose an autoimmune disease described in this section and discuss how it may affect an individual.

How can MHC proteins cause a problem during organ transplants?

AP CHAPTER SUMMARY

Summarize It

Compare and contrast passive and active immunity.

What is a consequence to the organism if helper T cells are destroyed by a pathogen, such as HIV?

An individual who has had chickenpox or has been vaccinated against the disease will rarely contract the disease again if exposed to it a second time. Explain why this might be.

FOLLOWING *the* BIG IDEAS

AP Essential Knowledge	Chapter Section
BIG IDEA 2 **2.A.3** Organisms must exchange matter with the environment to grow, reproduce and maintain organization.	34.1, 34.3, 34.4
2.B.2 Growth and dynamic homeostasis are maintained by the constant movement of molecules across membranes.	34.4
2.C.2 Organisms respond to changes in their external environments.	34.4
2.D.2 Homeostatic mechanisms reflect both common ancestry and divergence due to adaptation in different environments.	34.1, 34.2, 34.4
BIG IDEA 4 **4.A.4** Organisms exhibit complex properties due to interactions between their constituent parts.	34.2, 34.3
4.B.2 Cooperative interactions within organisms promote efficiency in the use of energy and matter.	34.2, 34.3

CHAPTER OVERVIEW

The digestive system breaks down food, allows for the absorption of nutrients, and eliminates undigestible remains. Compartmentalization, specialization, and increased surface area of organs are common themes seen in the evolution of the digestion tracts. This allows for more efficient removal of nutritional components of food.

34.1 Digestive Tracts

Essential Knowledge covered
2.A.3: *Organisms must exchange matter with the environment to grow, reproduce, and maintain organization.*
2.D.2: *Homeostatic mechanisms reflect both common ancestry and divergence due to adaptation in different environments.*

Recall It

Almost all animals have some sort of digestive tract for the ingestion and breakdown of food. Some animas have an incomplete digestive tract, while most have complete digestive tracts. Sponges are unique in that they have no digestive tract, but they do remove food particles from their environment through specialized cells in the inner lining of the organism. Animals acquire nutrients either continuously or through discontinuous feeds. Animals have adapted specialized diets as well which are often reflected in the structure of their dentition and digestive organs.

Review It

List four functions of a digestive system.

Describe a simple digestive tract.

34.1 Digestive Tracts *continued*

Describe the difference between an incomplete digestive tract and a complete digestive tract.

Use It

What is a rumen, and how does it aid in digestion in cattle?

Describe two ways an organism may show adaptation to their diet.

34.2 The Human Digestive System

Essential Knowledge covered
2.D.2: Homeostatic mechanisms reflect both common ancestry and divergence due to adaptation in different environments.
4.A.4: Organisms exhibit complex properties due to interactions between their constituent parts.
4.B.2: Cooperative interactions within organisms promote efficiency in the use of energy and matter.

Recall It

The complete digestive tract of the human contains many organs and accessory organs. Digestion takes place in two stages: mechanical digestion and chemical digestion. Food enters the mouth and moves through a series of passageways to the stomach. Digested food, called chime, then moves to the small intestine and then the large intestine. Digestive waste leaves the body through the anus. Along the way, many specialized organs and accessory organs play a role in providing digestive enzymes to aid in the breakdown of particular macromolecules into smaller molecules.

Review It

List three accessory organs of the human digestive system.

Using a diagram, draw the major structures and accessory organs of the human digestive tract in the correct order.

Use It

Using your diagram above, explain how a piece of pizza is mechanically and chemically digested by a human.

Describe the role of the large intestine.

How does the liver play a role in digestion?

Describe how the structure of the small intestine results in increased surface area for nutrient absorption.

34.3 Digestive Enzymes

Essential Knowledge covered
2.A.3: Organisms must exchange matter with the environment to grow, reproduce, and maintain organization.
4.A.4: Organisms exhibit complex properties due to interactions between their constituent parts.
4.B.2: Cooperative interactions within organisms promote efficiency in the use of energy and matter.

Recall It

Digestive enzymes play an important role in the breakdown of macromolecules as they pass through the digestive tract. There are many enzymes, and each is specialized to catalyze the breakdown of a particular macromolecule. The major components of food: carbohydrates, proteins, nucleic acids, and fats, each have a particular set of enzymes associated with their breakdown. Recall that enzymes have optimum conditions that they work best in, so particular enzymes are only found in specific regions of the human body.

Review It

Identify the digestive enzymes:

Description	Enzyme
breaks down proteins into peptides	
aids in the conversion of maltose to glucose	
the first enzyme to breakdown starch	
found in pancreatic juice and breaks down protein	
breaks down peptides into amino acids	
found in pancreatic juice and breaks down starch	
found in pancreatic juice and breaks down fat	

Use It

What are the major nutritional components of food which digestive enzymes break down?

Describe what enzymes break down the piece of pizza you wrote about in Section 34.2.

34.4 Nutrition and Human Health

Essential Knowledge covered
2.A.3: Organisms must exchange matter with the environment to grow, reproduce, and maintain organization.
2.B.2: Growth and dynamic homeostasis are maintained by the constant movement of molecules across membranes.

Recall It

Humans require a balanced diet of carbohydrates, lipids, proteins, and vitamins and minerals. Carbohydrates include sugar, starch, and fiber. Proteins are digested into amino acids. Lipids supply energy for cells but also are stored for the long term in the body as fat. Consumption of excess calories relative to the calories expended can lead to obesity. Obesity can cause a host of medical problems, including Type 2 diabetes.

34.4 Nutrition and Human Health *continued*

Review It

Identify the dietary component based on the description:

Description	Component
fats and oils which supply energy for cells	
twenty elements needed for various physiological functions, such as fluid balance	
eight molecules which adults cannot synthesizes but need for cellular processes	
organic compounds which regulate metabolic activities and are often part of coenzymes	
indigestible carbohydrates derived from plants	

Describe and differentiate between type 1 and type 2 diabetes.

Use It

Why is it suggested that people eat unsaturated fats in greater quantity than saturated fats?

Why is it important to maintain a healthy diet?

AP CHAPTER SUMMARY

Summarize It

What special features of the organs of the digestive system and interactions among these organs promote efficiency in the breakdown of food and absorption of nutrients?

How does the digestive system contribute to homeostasis?

FOLLOWING *the* BIG IDEAS

AP Essential Knowledge	Chapter Section
BIG IDEA 2 **2.A.3** Organisms must exchange matter with the environment to grow, reproduce and maintain organization.	35.1
2.C.1 Organisms use feedback mechanisms to maintain their internal environments and respond to external environmental changes.	35.3
2.D.2 Homeostatic mechanisms reflect both common ancestry and divergence due to adaptation in different environments.	35.1, 35.2
2.D.3 Biological systems are affected by disruptions to their dynamic homeostasis.	35.3
BIG IDEA 4 **4.A.4** Organisms exhibit complex properties due to interactions between their constituent parts.	35.1, 35.2
4.B.2 Cooperative interactions within organisms promote efficiency in the use of energy and matter.	35.1, 35.2
4.B.3 Interactions between and within populations influence patterns of species distribution and abundance.	35.3

CHAPTER OVERVIEW

Gas exchange between the body and the environment is the process of respiration. The evolution of respiration in animals can be traced from the gills of fish and tracheal system of insects to the vertebrate lung. In more advanced terrestrial vertebrates, there is cooperation between the respiratory and circulatory systems.

35.1 Gas-Exchange Surfaces

Essential Knowledge covered
2.A.3: *Organisms must exchange matter with the environment to grow, reproduce, and maintain organization.*
2.D.2: *Homeostatic mechanisms reflect both common ancestry and divergence due to adaptation in different environments.*
4.A.4: *Organisms exhibit complex properties due to interactions between their constituent parts.*
4.B.2: *Cooperative interactions within organisms promote efficiency in the use of energy and matter.*

Recall It

Respiration occurs in a series of steps in terrestrial animals: (1) ventilation, (2) external respiration, and (3) internal respiration. The ultimate goal of respiration is gas exchange with an animal's environment. Gas exchange occurs in a moist, thin, and large region of an animal's body. In aquatic invertebrates and vertebrates, this surface is the gills. Hydras use the entire body surface for gas exchange, as do earthworms. Insects have a network of trachea inside their body. Terrestrial vertebrates have lungs.

Review It

List three properties gas-exchange regions require in an organism in order for external respiration to be effective.

35.1 Gas-Exchange Surfaces *continued*

Identify the part or function of the respiratory system:

Description	Part/Function
organ that terrestrial vertebrates use to obtain oxygen	
organ found in aquatic organisms for exchanging gases in water	
gas exchange between a body's cells and the environment	
system of air tubes in insects	
process used by fish to transfer oxygen from the water into their blood	
air pockets within the lungs	

Use It

Compare and contrast the bronchioles in humans to the tracheoles in insects.

35.2 Breathing and Transport of Gases

Essential Knowledge covered
2.D.2: Homeostatic mechanisms reflect both common ancestry and divergence due to adaptation in different environments.
4.A.4: Organisms exhibit complex properties due to interactions between their constituent parts.
4.B.2: Cooperative interactions within organisms promote efficiency in the use of energy and matter.

Recall It

Inspiration (moving air into) and expiration (moving out of) the respiratory tract is the process of breathing. During inspiration, the volume of the lung increases and the air pressure decreases. During expiration, increased air pressure causes air in the lungs to move out. Most vertebrates use a tidal ventilation mechanism. Birds, however, use a one-way ventilation system, greatly improving their gas-exchange efficiency. Gas exchange involves both external respiration, as well as internal respiration. External respiration occurs in the lungs, where carbon dioxide leaves the blood and oxygen enters. Internal respiration occurs in the tissues, where oxygen leaves the blood and carbon dioxide enters.

Review It

Identify a similarity and one difference in respiration between reptiles, birds, and mammals.

35.2 Breathing and Transport of Gases *continued*

Given the definition on the left, provide the correct word associated with breathing and gas transport:

Definition	Word
an iron-containing group found in hemoglobin	
the act of moving air out of the lungs	
a horizontal muscles that divides the thoracic cavity and the abdominal cavity	
the act of moving air into of the lungs	
the amount of pressure a gas exerts	

Use It

Identify which occurs during external respiration and which occurs during internal respiration:

Process	Internal/External Respiration
Oxyhemoglobin gives up oxygen.	
Oxygen binds with hemogoblin.	
Carbon dioxide enters red blood cells and becomes carbaminohemoglobin or a bicarbonate ion.	
Carbonic acid is broken down with the aid of carbonic andhydrase into carbon dioxide and water.	

Draw a diagram illustrating the movement of carbon dioxide and oxygen through the respiratory and circulatory pathways in a human.

Using your diagram above, describe how respiration occurs in the human body.

35.3 Respiration and Human Health

Essential Knowledge covered
2.D.3: Biological systems are affected by disruptions to their dynamic homeostasis.
4.B.3: Interactions between and within populations influence patterns of species distribution and abundance.

Recall It

There are several disorders that affect the upper and lower respiratory tract in humans. They can be caused by allergies, infections, a genetic defect, or a toxin exposure. The upper respiratory tract includes the nasal cavities, sinuses, pharynx, and larynx. Many of the disorders that affect the upper respiratory tract are the result of pathogens and materials that are first filtered out by the nose. The lower respiratory tract includes the trachea, bronchi, and lungs. Disorders that affect the lower respiratory tract include choking, bronchitis, asthma, to pneumonia, lung cancer, and cystic fibrosis.

Review It

Identify the respiratory disease based on its description.

Description	Respiratory Disease
a genetic lung disease in which a faulty transport protein causes sticky mucus secretions that interfere with breathing	
a bacterial, viral, or fungal infection of the lungs which fill the bronchi or alveoli with pus or fluid	
a chronic and incurable lung disorder where alveoli are distended and damaged	
a tumor in the lungs	
the inflammation of the pharynx caused by a virus	

Use It

Why is there no vaccine for the common cold?

Summarize It

How have terrestrial vertebrates evolved to obtain oxygen?

How do the respiratory and circulatory systems work together to supply all cells of the body with oxygen and eliminate carbon dioxide?

FOLLOWING *the* BIG IDEAS

	AP Essential Knowledge	Chapter Section
BIG IDEA **2**	**2.C.1** Organisms use feedback mechanisms to maintain their internal environments and respond to external environmental changes.	36.2
	2.D.2 Homeostatic mechanisms reflect both common ancestry and divergence due to adaptation in different environments.	36.1
	2.D.3 Biological systems are affected by disruptions to their dynamic homeostasis.	36.2
BIG IDEA **4**	**4.A.4** Organisms exhibit complex properties due to interactions between their constituent parts.	36.1, 36.2
	4.B.2 Cooperative interactions within organisms promote efficiency in the use of energy and matter.	36.1, 36.2

CHAPTER OVERVIEW

All animals maintain their normal water-salt balance while excreting metabolic wastes and regulating their pH. The mechanisms in which animals regulate their body fluids is heavily dependent on the environment in which they live. In humans, the excretory system works together with the circulatory system to eliminate wastes while maintaining homeostasis.

36.1 Animal Excretory Systems

Essential Knowledge covered
2.D.2: Homeostatic mechanisms reflect both common ancestry and divergence due to adaptation in different environments.
4.A.4: Organisms exhibit complex properties due to interactions between their constituent parts.
4.B.2: Cooperative interactions within organisms promote efficiency in the use of energy and matter.

Recall It

Osmoregulation is the portion of homeostasis which balances the levels of water and salts in the body. Excretion is the removal of metabolic wastes. Animals excrete some form of nitrogenous wastes; the amino groups from amino acids that are not used to create energy. Nitrogenous wastes include urea, uric acid, and ammonia. Excretory organs among invertebrates and vertebrates vary greatly in complexity and structure. The different mechanisms of osmoregulation between aquatic vertebrates in marine and freshwater systems highlight evolutionary adaptations to maintain homoeostasis in different environments. The development of the kidney is an important adaptation that allows animals to survive on land.

Review It

Define *osmoregulation*.

36.1 Animal Excretory Systems *continued*

Using the flow charts below, identify the nitrogenous waste excreted by the following organisms, and draw an arrow pointing from the compound that requires the most energy to synthesize to the compound that requires the least energy.

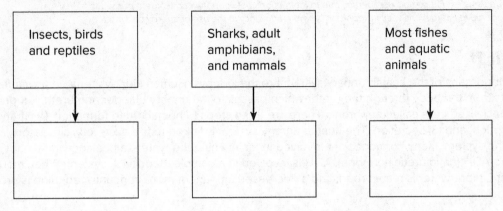

Use It

How do the kidneys help maintain homeostasis?

How have marine mammals and sea birds adapted to living in a high salt environment?

Which drinks more water, a bony freshwater fish or a marine bony fish, and why?

36.2 The Human Urinary System

Essential Knowledge covered
2.C.1: Organisms use feedback mechanisms to maintain their internal environments and respond to external environmental change.
2.D.3: Biological systems are affected by disruptions to their dynamic homeostasis.
4.A.4: Organisms exhibit complex properties due to interactions between their constituent parts.
4.B.2: Cooperative interactions within organisms promote efficiency in the use of energy and matter.

Recall It

The major organs of the human urinary system are the kidneys. Human kidneys remove unwanted products from the body through the ureter which passes to the urinary bladder and urethra. Kidneys have specialized cells called nephrons. There are three steps to human urine formation: (1) filtration, (2) reabsorption, and (3) secretion. The human urinary system is linked to the excretory, circulatory, and endocrine system. Many hormones are involved in maintaining the water-salt balance in humans, including ADH (antidiuretic hormone). ADH is involved in a complex feedback system. When more ADH is present, more water is reabsorbed, and a decreased amount of more concentrated urine is produced.

Review It

Place in order the structures which urine travels through in the human body: urethra, kidneys, ureter, urinary bladder.

List the three distinct processes of the urinary system.

Use It

Describe in detail the three processes of the urinary system.

glomerular filtration:

tubular reabsorption:

tubular secretion:

List four reasons the kidneys are the primary organs of homeostasis in humans.

If a person becomes dehydrated, how does the body respond?

Summarize It

How is excretion important to osmoregulation?

How do the liver, kidneys, and adrenal cortex work together to control the regulation of salt in the body?

FOLLOWING *the* BIG IDEAS

	AP Essential Knowledge	Chapter Section
BIG IDEA 3	**3.D.1** Cell communication processes share common features that reflect a shared evolutionary history.	37.4
	3.D.2 Cells communicate with each other through direct contact with other cells or from a distance via chemical signaling.	37.2
	3.D.3 Signal transduction pathways link signal reception with cellular response.	37.2
	3.E.1 Individual scan act on information and communicate it to others.	37.4
	3.E.2 Animals have nervous systems that detect external and internal signals, transmit and integrate information, and produce responses.	37.1, 37.2, 37.3, 37.4
BIG IDEA 4	**4.A.4** Organisms exhibit complex properties due to interactions between their constituent parts.	37.4
	4.B.2 Cooperative interactions within organisms promote efficiency in the use of energy and matter.	37.1, 37.2, 37.3, 37.4

CHAPTER OVERVIEW

While the simplest multicellular animals lack a nervous system, the nervous system evolved to become critical to the workings of more complex animals. From the monitoring of internal and external conditions, the nervous system maintains homeostasis through nervous tissue; neurons and neuroglia. Humans have both a central nervous system and peripheral nervous system, and a specialized portion of the brain called the neocortex.

37.1 Evolution of the Nervous System

Essential Knowledge covered
3.E.2: Animals have nervous systems that detect external and internal signals, transmit and integrate information, and produce responses.
4.B.2: Cooperative interactions within organisms promote efficiency in the use of energy and matter.

Recall It

Almost all animals have some sort of nervous system, the exception being simple animals such as sponges. The evolution of the nervous system ranges from a simple nerve net in hydras to complex cephalization in humans. While the structure of nervous systems differ between animals, the functions are similar with the nervous system being in control of coordination of movement and regulation of homeostasis.

Review It

List three functions of the nervous system.

37.1 Evolution of the Nervous System *continued*

Given the definition on the left, provide the correct word associated with the evolution of the nervous system:

Definition	Word
a brain and spinal cord	
a simple nervous system organization in which neurons are in contact with one another and the cells of the body	
a concentration of nervous tissue in the anterior or head region	
all the nerves and ganglia that lie outside the central nervous system	
a cluster of neuron cell bodies	

Use It

Rank the organisms from least complex nervous system (1) to most complex (5) and describe the major characteristics of each.

Complexity	Organism	Nervous System
	Octopus	
	Cat	
	Hydra	
	Earthworm	
	Crab	

What is the neocortex, what are its functions, and how is it different in humans than in other mammals?

37.2 Nervous Tissue

Essential Knowledge covered
3.D.2: Cells communicate with each other through direct contact with other cells or from a distance via chemical signaling.
3.D.3: Signal transduction pathway link signal reception with cellular response.
3.E.2: Animals have nervous systems that detect external and internal signals, transmit and integrate information, and produce responses.
4.B.2: Cooperative interactions within organisms promote efficiency in the use of energy and matter.

Recall It

Nervous tissue is comprised of neurons and neuroglia. Neurons have three major parts: (1) a cell body, (2) dendrites, and (3) an axon. Axons are often covered with a myelin sheath. Neurons are described in terms of their shape and function. Neurons include: (1) motor neurons, (2) sensory neurons, and (3) interneurons. Neuroglia are classified likewise, and include (1) astrocytes, (2) microglia, (3) oligodendrocytes, and (4) Schwann cells, which all play a specific function within different areas of the nervous system. Nerve impulse or action potentials are conducted by the axons. An action potential occurs when there is a rapid change in polarity across the axon membrane. The sodium-potassium pump brings more sodium inside the axon than outside, and when a resting potential of +35 mV is reached, depolarization beings. Neurotransmitters are molecules that can move across a synaptic cleft and elicit an action potential.

Review It

List the three types of neurons.

Describe four neuroglia and their roles in the nervous system.

Use It

Using a diagram, show how a neurotransmitter transmits an action potential across a synapse.

Describe what is happening in the neuron illustrated below.

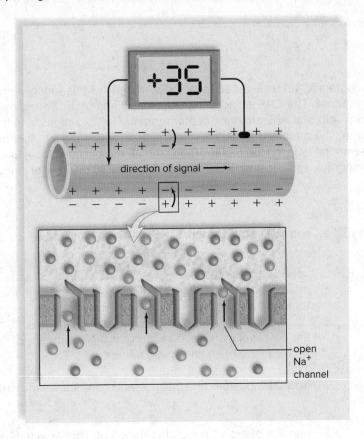

Describe the role the following neurotransmitters play in the body:

Neurotransmitters	Role
serotonin	
endorphins	
acetylcholine (ACh)	
norepinephrine	
dopamine	

37.3 The Central Nervous System

Essential Knowledge covered
3.E.2: Animals have nervous systems that detect external and internal signals, transmit and integrate information, and produce responses.
4.B.2: Cooperative interactions within organisms promote efficiency in the use of energy and matter.

Recall It

The central nervous system (CNS) receives sensory input, integrates this sensory information, and generates a motor response. The CNS includes the spinal cord and brain. The spinal cord is a bundle of nerves enclosed in the vertebral column, is the center for many reflex actions, and is a pathway for communication between the brain and spinal nerves. The brain is divided into four major lobes: (1) the frontal lobe, (2) the temporal lobe, (3) the parietal lobe, and (4) the occipital lobe. Each lobe functions to somehow capture, integrate, and determine the correct response to the sensory input coming in.

Review It

Identify the structure of the Central Nervous System (CNS) given its function:

Structure	Function
	maintains homeostasis by regulating hunger, sleep, thirst, and body temperature
	the center of many reflex actions as well as means of communication between the brain and spinal nerves
	maintains posture and balance and coordinates muscle actions
	bundles of axons that bridge the cerebellum and the rest of the CNS
	accounts for sensation, voluntary movement, and all processes required for memory, learning, and speech
	regulates the heartbeat, breathing, and blood pressure
	the gatekeeper for sensory information enroute to the cerebral cortex
	the two structures of the limbic system which are essential for learning and memory
	fills the spaces between the meninges to cushion and protect the CNS
	communicates and coordinates activities of the brain
	integrate motor commands

Use It

List the three specific functions of the central nervous system.

Describe what parts of the brain allow you to recall something from a long time ago.

Suppose you were walking in the woods and were startled by a snake slithering across the path. What part of the brain allowed you to be startled?

37.4 The Peripheral Nervous System

Essential Knowledge covered
3.D.1: Cell communication processes share common features that reflect a shared evolutionary history.
3.E.1: Individuals can act on information and communicate it to others.
3.E.2: Animals have nervous systems that detect external and internal signals, transmit and integrate information, and produce responses.
4.A.4: Organisms exhibit complex properties due to interactions between their constituent parts.
4.B.2: Cooperative interactions within organisms promote efficiency in the use of energy and matter.

Recall It

The peripheral nervous system (PNS) contains our nerves and is divided into the (1) somatic system and (2) autonomic system. The somatic system takes information from skin receptors and joins it to the CNS, and carries motor commands from the CNS to the skeletal muscles. The autonomic system regulates the activities of cardiac and smooth muscle, and the activity of glands. The autonomic system operates without our knowledge and is divided into sympathetic and parasympathetic responses.

Review It

List three types of nerves in the peripheral nervous system.

What type of reaction is illustrated in the diagram below?

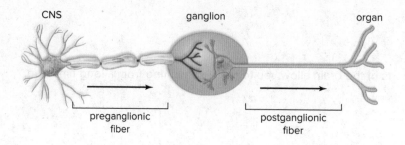

Use It

Compare and contrast the somatic motor and autonomic motor pathways.

What neurotransmitter is released in the sympathetic division of the PNS when someone is under attack and what does it do?

Summarize It

Describe the organization of the vertebrate brain.

Explain the difference between a neuron and a neurotransmitter.

Describe how your nervous system might respond if you were to be poked with a pin.

Following *the* Big Ideas

AP Essential Knowledge	Chapter Section
1.B.1 Organisms share many conserved core processes and features that evolved and are widely distributed among organisms today.	38.2, 38.3, 38.4
1.C.3 Populations of organisms continue to evolve.	38.1, 38.2, 38.3, 38.4

BIG IDEA 1

Chapter Overview

Animals have evolved many different ways to perceive information. These adaptations are known as sense organs. Sense organs work closely with the nervous system to coordinate responses from sensing change. There are many types of sensory receptors. While the intricacies of each sensory receptor and the associated organs are not required for you to know in detail for AP, there are many interesting examples which highlight the evolution of homeostasis in animals.

38.1 Sensory Receptors

Essential Knowledge covered
1.C.3: Populations of organisms continue to evolve.

Recall It

Selective pressures have resulted in the evolution of many different types of sensory receptors in animals. A sensory receptor converts a stimulus into an action potential. Types of sensory receptors include: chemoreceptors, electromagnetic receptors, photoreceptors, mechanoreceptors, and thermoreceptors. These receptors are classified by the source of stimuli received.

Review It

What is sensory transduction?

For the sensory receptors listed below, provide their function and an example of each.

Receptor	Function	Example
mechanoreceptors		
thermoreceptors		
electromagnetic receptors		
chemoreceptors		

38.1 Sensory Receptors *continued*

Use It

Why do animals have sensory receptors?

38.2 Chemical Senses

Essential Knowledge covered
1.B.1: Organisms share many conserved core processes and features that evolved and are widely distributed among
        organisms today.
1.C.3: Populations of organisms continue to evolve.

Recall It

The most primitive kind of sense receptor is thought to be chemoreception as it is found almost universally in animals. Receptors sensitive to certain chemicals can be found on many surfaces of an animal's body. The human tongue is an example of a chemoreceptor that contains ~3000 taste buds. Olfactory cells are located in the nose. The taste buds and olfactory cells frequently work together in the human body to send messages to the cerebral cortex.

Review It

Why is chemoreception thought to be the most primitive sense?

Use It

Compare and contrast the senses of taste and smell in humans.

How do chemoreceptors vary throughout the animal kingdom?

38.3 Sense of Vision

Essential Knowledge covered
1.B.1: Organisms share many conserved core processes and features that evolved and are widely distributed among organisms today.
1.C.3: Populations of organisms continue to evolve.

Recall It

Sensory receptors that are sensitive to light are called photoreceptors. Arthropods have compound eyes; whereas, vertebrates and squid have camera-type eyes. When eyes face forward, organisms have stereoscopic vision and panoramic vision. In the human eye, there are three layers, the sclera which becomes the cornea, conjunctiva vessels, and pigment that absorbs stray light rays. The iris regulates the size of the pupil. The pupil regulates the light entering the eye, and the images. Photoreceptors called rod cells and cone cells allow us to see light and color.

Review It

Provide one reason is it beneficial for a rabbit to have panoramic vision.

Why do cones provide us with a sharper image of an object than the rod cells?

Use It

Using the following illustration, describe how rhodopsin works.

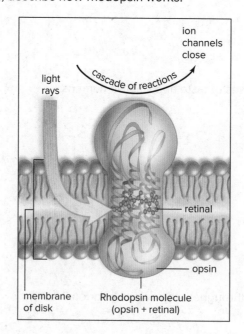

38.4 Sense of Hearing and Balance

Essential Knowledge covered
1.B.1: Organisms share many conserved core processes and features that evolved and are widely distributed among organisms today.
1.C.3: Populations of organisms continue to evolve.

Recall It

Animals use mechanoreception to sense physical contact on the skin or movement in the surrounding environment. Hearing also is a type of mechanoreception, as it stimulated by the vibration of sound waves. Hearing in humans requires the ear, the cochlear nerve, and the auditory areas of the cerebral cortex. The ear also contains receptors for our sense of equilibrium.

Review It

Describe the following components associated with mechanoreception:

Component	Description
middle ear	
gravitational equilibrium	
the organ of Corti	
inner ear	
rotational equilibrium	
outer ear	
auditory tube	
lateral line	

Use It

Compare and contrast the lateral line in fish and the ear in humans.

Describe what happens inside your ear when you shake your head 'no' versus nod your head 'yes'.

38.5 Somatic Senses

Essential Knowledge covered
1.B.1: Organisms share many conserved core processes and features that evolved and are widely distributed among organisms today.

Recall It

Somatic senses are receptors associated with the skin, muscles, joints, and viscera. Proprioceptors are mechanoreceptors involved in reflex actions that maintain the body's equilibrium and posture. Cutaneous receptors are receptors found in the skin which are sensitive to touch and temperature. Skin, internal organs, and tissue have pain receptors.

Review It

Identify the function and provide an example of each of the somatic senses below:

Component	Function	Example
Proprioceptors		
Cutaneous receptors		
Pain receptors		

Use It

How do the somatic senses work?

AP CHAPTER SUMMARY

Summarize It

Compare and contrast photoreceptors in planarians, dragonflies, and primates.

What is an example of a sensory receptor, and what features allow it to receive information from the environment?

Following *the* Big Ideas

AP Essential Knowledge	Chapter Section
BIG IDEA 4 **4.A.2** The structure and function of subcellular components, and their interactions, provide essential cellular processes.	39.2
4.A.4 Organisms exhibit complex properties due to interactions between their constituent parts.	39.1, 39.2, 39.3
4.B.2 Cooperative interactions within organisms promote efficiency in the use of energy and matter.	39.1, 39.3

Chapter Overview

Skeletons and muscles provide the support systems for the body of animals. Skeletons provide rigidity, protection, and surfaces for muscle attachment. Muscles are composed of contractive tissue. The contraction of muscles is dependent on their close relationship with the nervous system. The ability for a body to respond to movement is critical for its survival.

39.1 Diversity of Skeletons

Essential Knowledge covered
4.A.4: Organisms exhibit complex properties due to interactions between their constituent parts.
4.B.2: Cooperative interactions within organisms promote efficiency in the use of energy and matter.

Recall It

There are many types of skeletons in the animal kingdom. All skeletons, however, provide a support system for animals. Hydrostatic skeletons are fluid-filled gastrovascular cavities that utilize fluid pressure to offer support and resistance to the contraction of muscles. Some animals have exoskeletons; skeletons that are external coverings compromised of a stiff material. Echinoderms and vertebrates have endoskeletons. Endoskeletons are made up of internal rigid structures.

Review It

What is the function of a skeletal system?

39.1 Diversity of Skeletons *continued*

Describe the three types of skeletal systems listed below.

Skeleton	Description
hydrostatic skeleton	
endoskeleton	
exoskeleton	

Use It

Describe four skeletal structures of mammals that are adapted to a particular mode of locomotion.

List three advantages of a jointed endoskeleton.

39.2 The Human Skeletal System

Essential Knowledge covered
4.A.2: The structure and function of subcellular components, and their interactions, provide essential cellular processes.
4.A.4: Organisms exhibit complex properties due to interactions between their constituent parts.

Recall It

The human skeleton has five major functions. First, it offers support of the body. Second, it protects our vital internal organs. It also provides sites for muscles to attach. Fourth, it becomes a storage reservoir for ions, including calcium and phosphate. Finally, the skeleton aids in the production of blood cells. There are many different bones in the human skeletal system, the memorization of which are beyond the scope of this AP course.

Review It

List two function of the vertebral column.

Describe three ways in which the skeletal system supports homeostasis.

Copyright © McGraw-Hill Education

Use It
Explain the roles of osteoblasts, osteoclasts, and osteocytes in bone growth and renewal.

39.3 The Muscular System

Essential Knowledge covered
4.A.4: Organisms exhibit complex properties due to interactions between their constituent parts.
4.B.2: Cooperative interactions within organisms promote efficiency in the use of energy and matter.

Recall It

Skeletal muscles are attached to the skeleton by tendons. When a muscle is at rest, only some fo the fibers in the muscles are contracting, and the muscle exhibits tone. Maximum sustained muscle contraction is known as tetany. Muscle fibers are divided into contractile sarcomeres which are made up of myofibrils. In the sliding filament model, thin actin filaments slide past thick myosin filaments, creating a muscle contraction. Calcium ions, stored in the sarcoplasmic reticulum, are essential in the binding of myosin to actin. ATP supplies the energy for muscle contraction and is generated through creatine phosphate breakdown and fermentation during oxygen debt. All muscle contractions are initiated when a nerve impulse travels down from a motor neuron to a neuromuscular junction, causing acetylcholine to bind to the sarcolemma.

Review It

Describe the state of the muscle in the illustration below. Identify the actin, myosin, and sarcomere.

What determines the duration of a muscle contraction?

39.3 The Muscular System *continued*

Use It

How does a nerve impulse translate into a muscle contraction? Be sure to identify the neurotransmitter, and the role of calcium and ATP in your answer.

AP CHAPTER SUMMARY

Summarize It

What is the relationship between the ability of a muscle to contract and input from the nervous system?

What advantages do vertebrates gain by having an endoskeleton as opposed to a hydroskeleton?

40 Hormones and Endocrine Systems

FOLLOWING *the* BIG IDEAS

AP Essential Knowledge	Chapter Section
BIG IDEA 2 **2.C.1** Organisms use feedback mechanisms to maintain their internal environments and respond to external environmental changes.	40.2, 40.3
2.C.2 Organisms respond to changes in their external environments.	40.3
2.D.2 Homeostatic mechanisms reflect both common ancestry and divergence due to adaptation in different environments.	40.3
2.D.3 Biological systems are affected by disruptions to their dynamic homeostasis.	40.2
2.E.2 Timing and coordination of physiological events are regulated by multiple mechanisms.	40.1, 40.2, 40.3
BIG IDEA 3 **3.D.1** Cell communication processes share common features that reflect a shared evolutionary history.	40.1, 40.3
3.D.2 Cells communicate with each other through direct contact with other cells or from a distance via chemical signaling.	40.2
3.D.3 Signal transduction pathways link signal reception with cellular response.	40.1
3.D.4 Changes in signal transduction pathways can alter cellular response.	40.3
3.E.2 Animals have nervous systems that detect external and internal signals, transmit and integrate information, and produce responses.	40.2

CHAPTER OVERVIEW

Secreted from glands scattered throughout the body, the endocrine system produces hormones to coordinate the activities of the body's other organ systems. Hormones are fundamental in helping the body maintain homeostasis. Both negative and feedback mechanisms regulate the production of hormones in the body. Hormones work at a much slower rate than neurotransmitters.

40.1 Animal Hormones

Essential Knowledge covered
2.E.2: Timing and coordination of physiological events are regulated by multiple mechanisms.
3.D.1: Cell communication processes share common features that reflect a shared evolutionary history.
3.D.3: Signal transduction pathway link signal reception with cellular response.

Recall It

The endocrine system secretes chemical signals known as hormones. Hormones are carried to target cells throughout the body. It takes time for hormones to be delivered to the target cells. Once the hormones are delivered, however, there is a prolonged response. Exocrine glands secrete their products into ducts, and endocrine glands secrete their products into the bloodstream. There are many types of hormones, which exert a wide range of effects on cells. Hormones pass messages between cells, body parts, and individuals. Pheromones influence the behavior of others. Hormones derived from cholesterol are called steroids, and hormones that are peptides, proteins, and other modified amino acids are peptide hormones. A hormone that never enters a cell but delivers the signal that a cascade of metabolic activity needs to occur is called the first messenger, and the molecule that receives this message and sets o the cascade is called the second messenger.

40.1 Animal Hormones *continued*

Review It

List four major endocrine glands in the human body, the hormone they secrete, and the function the hormone plays.

What is cAMP? What does it do?

Use It

Describe two ways humans may be influenced by pheromones.

Compare and contrast how a peptide hormone and a steroid hormone activate a change in the body.

40.2 Hypothalamus and Pituitary Gland

Essential Knowledge covered
2.C.1: Organisms use feedback mechanisms to maintain their internal environments and respond to external environmental change.
2.D.3: Biological systems are affected by disruptions to their dynamic homeostasis.
2.E.2: Timing and coordination of physiological events are regulated by multiple mechanisms.
3.D.2: Cells communicate with each other through direct contact with other cells or from a distance via chemical signaling.
3.D.4: Changes in signal transduction pathways can alter cellular response.

Recall It

The hypothalamus regulates the body's internal environment. It influences the heartbeat, blood pressure, appetite, body temperature, and water balance. It also controls the secretions of the pituitary gland. The pituitary gland is divided into the posterior pituitary and the anterior pituitary. The posterior pituitary stores and secretes two hormones, ADH and oxytocin, which are produced by the hypothalamus. The hypothalamus controls the secretions of the anterior pituitary, and the anterior pituitary controls the secretions of the thyroid, adrenal cortex, and gonads.

Review It

Explain how the hypothalamus and pituitary gland work together.

Use It

This figure illustrates the pathway of how TSH is produced in the thyroid gland. Explain what is happening in the illustration.

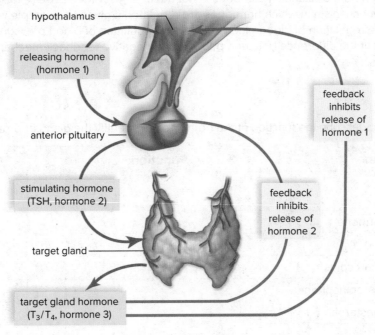

Explain the difference between a positive and negative feedback system in hormone production and give an example of each.

Describe the effects of the production of too little versus too much human growth hormone (GH) during childhood.

40.3 Other Endocrine Glands and Hormones

Essential Knowledge covered

2.C.1: *Organisms use feedback mechanisms to maintain their internal environments and respond to external environmental change.*

2.C.2: *Organisms respond to changes in their external environment.*

2.D.2: *Homeostatic mechanisms reflect both common ancestry and divergence due to adaptation in different environments.*

2.E.2: *Timing and coordination of physiological events are regulated by multiple mechanisms.*

3.D.1: *Cell communication processes share common features that reflect a shared evolutionary history.*

3.D.4: *Changes in signal transduction pathways can alter cellular response.*

Recall It

The thyroid and parathyroid glands, adrenal glands, pancreas, pineal gland, thymus, and other tissues produce hormones secondarily. There are many different types of hormones in the human body. The memorization of each hormone and what it does is not necessary for the scope of this AP course, but understanding the mechanism in which hormones are regulated if given an example is. This includes example such as the regulation of blood glucose levels or regulation of blood pressure and volume. Take a second look at the difference between positive and negative feedback mechanisms as described in previous chapters.

Review It

Using your textbook, describe the function of the hormone and identify the gland or organ it comes from.

Hormone	Function	Gland/Organ
aldosterone		
atrial natriuretic hormone		
calcitonin		
cortisol and glucocorticoids		
epinephrine and norepinephrine		
estrogen and progesterone		
insulin		
leptin		
melatonin		
parathyroid hormone		
prostagladin		
testosterone and androgens		
thyroxine		

Use It

Describe how calcitonin and PTH blood calcium levels.

Describe how insulin and glucagon regulate blood glucose levels.

Using the following figure, describe the level of melatonin production in relationship to the amount of light illustrated in each of the graphs on the left. Melatonin is depicted by the blue line, the yellow represents the amount of light, and the gray represents darkness.

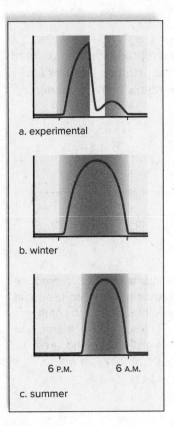

a. experimental

b. winter

c. summer

6 P.M.   6 A.M.

AP CHAPTER SUMMARY

Summarize It

How does negative and positive feedback regulate hormone production and activity? Pick one example of each from this chapter to describe in a paragraph below.

Answers will vary. A negative feedback example is ADH release by the pituitary to reabsorb.

FOLLOWING *the* BIG IDEAS

AP Essential Knowledge	Chapter Section
BIG IDEA 2 **2.C.1** Organisms use feedback mechanisms to maintain their internal environments and respond to external environmental changes.	41.1, 41.2, 41.3, 41.4, 41.5
2.E.2 Timing and coordination of physiological events are regulated by multiple mechanisms.	41.1

CHAPTER OVERVIEW

Animals reproduce through either sexual or asexual reproduction. Sexual reproduction, as covered in Chapter 10, involves gametes and produces genetically unique offspring. Human gametes are produced in different ways by males and females. Both male and female reproductive systems utilize hormone-based feedback mechanisms in gamete production.

41.1 How Animals Reproduce

Essential Knowledge covered
2.C.1: *Organisms use feedback mechanisms to maintain their internal environments and respond to external environmental changes.*
2.E.2: *Timing and coordination of physiological events are regulated by multiple mechanisms.*

Recall It

Animals reproduce either asexually or sexually. As covered in Chapter 9, asexual reproduction produces genetically identical offspring from one parent. As covered in Chapter 10, sexual reproduction produces genetically dissimilar offspring. Most animals that reproduce sexually are dioecious, although some monoecious species exist. Gametes are produced in organs called gonads. Testes produce sperm, and ovaries produce eggs. Eggs and sperm are derived from germ cells. Life strategist that can be seem within the animal kingdom include: oviparous, ovoviviparous, and viviparous; depending on where fertilized eggs are deposited as they develop.

Review It

Provide a definition and an example of an organism for the following life history strategies.

Life History Strategy	Definition	Example Organism
oviparous		
ovoviviparous		
viviparous		

Use It

While most animals are dioecious, some coral fish exhibit sequential hermaphroditism. Describe what happens with these coral reef fish.

41.1 How Animals Reproduce *continued*

Provide one advantage and one disadvantage for asexual reproduction in animals.

41.2 Human Male Reproductive System

Essential Knowledge covered
2.C.1: Organisms use feedback mechanisms to maintain their internal environments and respond to external environmental change.

Recall It

The human male reproductive system contains the testes, epididymitis, vasa deferentia, seminal vesicles, prostate gland, urethra, and penis. The testes produce sperm. The seminal vesicles, prostate gland, and bulbourethal gland provide a fluid medium for the sperm, which move from the vas deferens through the ejaculatory duct to the urethra in the penis. Sperm is produced through a feedback loop which involves the hormones GnRH, FSH, and LH. GnRH stimulates the anterior pituitary to produce FSH and LH. FSH stimulates the testes to produce sperm, and LH stimulates the testes to produce testosterone. Testosterone and inhibin exert negative feedback control over the hypothalamus and anterior pituitary.

Review It

Fill in the function or the component missing in the chart of the human male reproductive system.

Component	Function
vas deferens	
	primary male sex organs
	conducts sperm and urine
prostate gland	
epididymis	
	provides large surface area for sperm development
	the main sex hormone in males
semen	
seminal vesicles	
	male gametes

List the three parts of a mature sperm and their corresponding functions.

41.2 Human Male Reproductive System *continued*

Use It

Draw a diagram showing how the hypothalamus is involved in the production of sperm, and explain the feedback mechanism.

41.3 Human Female Reproductive System

Essential Knowledge covered
2.C.1: Organisms use feedback mechanisms to maintain their internal environments and respond to external environmental change.

Recall It

The human female reproductive system contains the vagina, uterus, uterine tubes, and ovaries. The vagina is the birth canal and organ of sexual intercourse. Fertilization may occur in the uterine tube and development occurs in the uterus. The ovaries produce one oocyte per month in what is known as the ovarian cycle. During the ovarian cycle, one follicle matures into an oocyte that can be fertilized and produces a secondary oocyte that becomes the corpus luteum. The follicle and corpus luteum produce estrogen and progesterone; whereas, the anterior pituitary produces FSH and LH. The menstrual cycle occurs with the ovarian cycle. Feedback control of the hypothalamus and anterior pituitary causes the levels of estrogen and progesterone to fluctuate during the cycles. When these hormones are at a low level, menstruation begins.

Review It

Fill in the function or the component missing in the chart of the human female reproductive system.

Component	Function
	the female sex hormones
oocyte	
	the birth canal
	houses developing embryo
follicle	
	the uterine lining where an embryo becomes implanted
corpus luteum	
	primary female sex organs

41.3 Human Female Reproductive System *continued*

Use It

Using the diagram below, describe how the ovarian cycle is driven by hormone levels, and how this drives the uterine cycle.

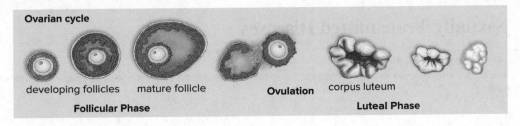

Draw a diagram showing how the hypothalamus is involved in the production of oocytes.

41.4 Control of Human Reproduction

Essential Knowledge covered
2.C.1: Organisms use feedback mechanisms to maintain their internal environments and respond to external environmental changes.

Recall It

Human reproductive contraception and technologies are numerous. Contraceptive devices and medications are used to reduce the chance of pregnancy. This includes everything from mechanical barriers to hormone based interruption of a pregnancy. Human reproductive technologies allow infertile couples to achieve pregnancy. There are many processes that can be used to achieve, as well.

Review It

List four methods of birth control.

Describe two reproductive technologies.

41.4 Control of Human Reproduction *continued*

Chose one form of hormonal birth control, and explain how it influences the human reproductive system.

41.5 Sexually Transmitted Diseases

Essential Knowledge covered
2.C.1: Organisms use feedback mechanisms to maintain their internal environments and respond to external environmental changes.

Recall It

Human disease is frequently spread through sexual intercourse. There are both viral and bacterial sexually transmitted diseases (STDs). STDs caused by viruses include AIDS, genital herpes, genital warts, cervical cancer, and hepatitis A and B. Major bacterial STDs include chlamydia and gonorrhea. The effects of STDs range from inflammation to warts, to more severe problems such as neurological and autoimmune disorders. Birth control, such as condoms and spermicide, only offer some protection against STDs.

Review It

List three viral sexual transmitted diseases.

List three bacterial sexual transmitted diseases.

Use It

What is highly active antiretroviral therapy and why has it become effective in the fight against the spread of AIDS?

Summarize It

What is the relationship between the hypothalamus/pituitary and the male and female reproductive systems?

How did shelled eggs allow for animals to colonize land?

FOLLOWING *the* BIG IDEAS

	AP Essential Knowledge	Chapter Section
BIG IDEA 2	**2.E.1** Timing and coordination of specific events are necessary for the normal development of an organisms, and these events are regulated by a variety of mechanisms.	42.1, 42.2, 42.3
BIG IDEA 3	**3.B.1** Gene regulation results in differential gene expression, leading to cell specialization.	42.2
	3.B.2 A variety of intercellular and intracellular signal transmissions mediate gene expression.	42.2
	3.D.2 Cells communicate with each other through direct contact with other cells or from a distance via chemical signaling.	42.1

CHAPTER OVERVIEW

Following fertilization, a diploid zygote begins the journey of development into an animal. There are many different stages of development that proceed from cellular to tissue to organ stages. Development is a perfectly timed process. Differential gene expression and tissue-specific proteins result in cell specialization. Homeotic genes influence development through determining the number and type of segments in an animal's body. Apoptosis, or cell death, also plays a major role in development.

42.1 Early Developmental Stages

Essential Knowledge covered
2.E.1: Timing and coordination of specific events are necessary for the normal development of an organism, and these events are regulated by a variety of mechanisms.
3.D.2: Cells communicate with each other through direct contact with other cells or from a distance via chemical signaling.

Recall It

Following fertilization, embryonic development occurs. Embryonic development occurs in three major stages: (1) cellular stages, (2) tissue stages, and (3) organ stages. The cellular stages are cleavage and formation of the blastula. The tissue stage is the development of the gastrula, which forms three layers of cells: (1) the ectoderm, (2) the endoderm, and (3) the mesoderm. These three layers will then develop into adult organs. The development of organs is known as the organ stage of development.

Review It

Describe the events that unfold when a sperm makes its way through the corona radiata of the oocyte.

42.1 Early Developmental Stages *continued*

Identify the germ layer in which the following structures develop from.

Structure	Germ Layer
The nervous system	
The thyroid and parathryoid glands	
Tooth enamel	
The lining of the digestive tract	
The cardiovascular system	
Hair and nails	
The outerlayer of the digestive system	

Use It

Draw a diagram of how a zygote develops into a morula, a blastula, and then forms three germ layers.

42.2 Developmental Processes

Essential Knowledge covered
2.E.1: Timing and coordination of specific events are necessary for the normal development of an organism, and these events are regulated by a variety of mechanisms.
3.B.1: Gene regulation results in differential gene expression, leading to cell specialization.
3.B.2: A variety of intercellular and intracellular signal transmission mediate gene expression.

Recall It

Cellular differentiation and morphogenesis are key processes in development, coupled with growth. Cellular differentiation is the process in which cells become specialized. A zygote is totipotent, meaning it has the ability to generate the entire organism. Every cell in the body has the same genes. It is what genes are turned on or off which allow for cellular differentiation. Morphogenesis is the process that forms the shape and form of the body; which includes cell movement and pattern formation. Homeotic genes, or selector genes, select for segmental identity. These genes are highly conserved. Apoptosis, or programmed cell death, is also important morphogenesis.

Review It

Identify the term associated with description below.

Description	Term
the ability of a zygote to generate the entire organism	
a functionally important 60 amino-acid sequence in a homeobox	
the parceling out of maternal determinants during mitosis	
how tissues and organs are arranged in the body	
the ability for one embryonic tissue to influence the development of another	
a structural feature in a homeotic gene that all organisms share	
when cells become specialized in structure and function	
programmed cell death	
substances in the cytoplasm which influence the course of development	
a gene which selects for segmental identity	

Use It

Scientists classify the cytoplasm of a frog egg as polar. What does this mean, and why is this important to development?

Compare and contrast maternal determinants and induction.

Answer the following questions concerning the diagram below.

Mouse *HOX* genes

Fruit fly *HOX* genes

42.2 Developmental Processes *continued*

Which genes occur in the same order in the diagram above?

What is the importance of the homeodomain in the homeotic genes?

42.3 Human Embryonic and Fetal Development

Essential Knowledge covered
2.E.1: Timing and coordination of specific events are necessary for the normal development of an organism, and these events are regulated by a variety of mechanisms.

Recall It

Human development can be divided into embryonic development and fetal development. Extraembryonic membranes in humans include the chorion, the amnion, the allantois, and the yolk sac. The yolk sac provides nourishment, the allantois collects nitrogenous waste, the aminon protects the developing embryo, and the chorion carries out gas exchange.

Review It

Describe what occurs during the weeks of embryonic development. Use the words: *umbilical cord, HCG, blastocyst, placenta, implantation, trophoblast.*

Week	Development
1	
2	
3	
4 and 5	
6 through 8	

List the three stages of birth.

Use It

What is the importance of the placenta?

42.4 The Aging Process
Extending Knowledge

Recall It
Aging and death are a critical part of the evolutionary process. New genetic combinations continuously replace individuals that die. Aging has many effects on organ systems which often result in discomfort and disease. There are many theories on why we age. Some theories suggest aging is due to preprogrammed genetic events, while others suggest it is due to the accumulation of cellular damage.

Review It
Describe what would happen if animals did not age or die.

Identify an example of aging on two organ systems in the human body.

Identify the two major hypotheses concerning why we age.

AP CHAPTER SUMMARY

Summarize It
Compare and contrast the tissue stage of development between a frog and lancelet.

Explain the role of *HOX* genes in development.

Provide an example of how apoptosis is important during development.

UNIT SEVEN: COMPARTIVE ANIMAL BIOLOGY

Chapter 31 Animal Organization and Homeostasis • Chapter 32 Circulation and Cardiovascular Systems • Chapter 33 The Lymphatic and Immune Systems • Chapter 34 Digestive Systems and Nutrition • Chapter 35 Respiratory Systems • Chapter 36 Body Fluid Regulation and Excretory Systems • Chapter 37 Neurons and Nervous Systems • Chapter 38 Sense Organs • Chapter 39 Locomotion and Support Systems • Chapter 40 Hormones and Endocrine Systems • Chapter 41 Reproductive Systems • Chapter 42 Animal Development and Aging

Multiple Choice Questions

Directions: For each question or incomplete statement chose one of the four suggested answers or completions listed below.

Use the following diagram to answer questions 1 and 2.

The below diagram shows that when the genes *Sredbf-1* and *Srebf2* activate two micro RNAs miR-33a and miR-33b, the genes involved in the breakdown of fatty acid (CROT, CPT1a, and HADHB) and cholesterol transport (ABCA1, ABCG1, NPC1) are subsequently inhibited.

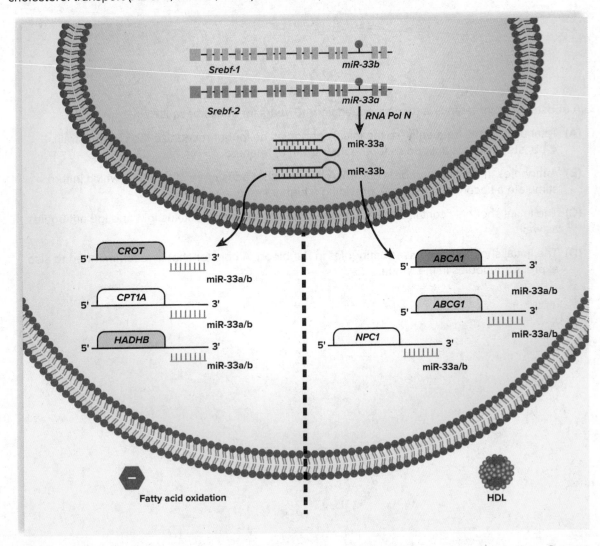

Data obtained from: Fernández-Hernando, Carlos, et al. 2011. MicroRNAs in lipid metabolism. *Current Opinion in Lipidology* 22 (2): 86.

1. Which homeostatic mechanism is most effected by the activation of these two miRNAs?

    (A) Temperature regulation

    (B) Blood glucose levels

    (C) Positive control

    (D) Lipid metabolism

2. Blocking the expression of miR-33a and miR-33b may lead to reductions in disease in which organ?

    (A) The heart

    (B) The liver

    (C) The gallbladder

    (D) All of the above

3. A booster shot for tetanus is recommended very 10 years for adults because

    (A) Tetanus mutates frequently, so the initial antibodies no longer recognize the current strain. A booster contains antibodies for the new strain of tetanus.

    (B) Antibodies in the first shot break down. A booster shot adds more antibodies, which in turn stimulate a secondary response, providing a higher immune response.

    (C) The initial shot only contains IgG antibodies. The booster shot contains IgM and IgE antibodies as well.

    (D) The initial shot only expresses antibodies in the blood. A booster shot is recommended to also express antibodies in the lymph.

4. Leatherback turtles have been observed to control their heat gain behaviorally through increasing their flipper stroke count when exposed to colder water, and also physiologically through regulation of blood flow. This would indicate that leatherback turtles are:

(A) Homeothermic

(B) Poikilothermic

(C) Ectotherms

(D) Thermostatic

5. The human reproductive system is regulated by various mechanisms. Which of the following demonstrates a positive feedback system?

(A) During the follicular phase of the menstrual cycle, estrogen is secreted and controls the secretion of FSH and LH.

(B) After conception, the placenta produces HCG which maintains the corpus luteum.

(C) During birth, the release of oxytocin causes increased uterine contractions.

(D) The human birth rate is influenced by birth weight.

6. Given the knowledge that marine fishes live in a hypertonic environment and must constantly drink seawater and excrete salt to make up for the loss of water through osmosis, which statement must be true about freshwater bony fishes?

(A) Freshwater bony fishes also live in hypertonic environments, lose water through osmosis, and drink a lot of water.

(B) Freshwater bony fishes gain water through osmosis, drink little, and discharge excess water in large quantities of urine.

(C) Freshwater bony fishes do not excrete salt through their gills.

(D) Freshwater bony fishes are isotonic to their environment.

Free Response Questions

Directions: Read the questions carefully and completely. Then, plan your answer and write your response in the space provide. Write your answer out in paragraph form.

1. Unlike the flu shot, a rabies shot is administered *after* exposure to the virus. **Explain** the science behind the timing of these two vaccines.

2. Myelinated axons have been found in both vertebrate and invertebrate groups. Myelin sheaths in some invertebrates are even similar in structure and function to vertebrates. Below is a simplified phylogeny showing some of the invertebrates found with myelin sheaths.

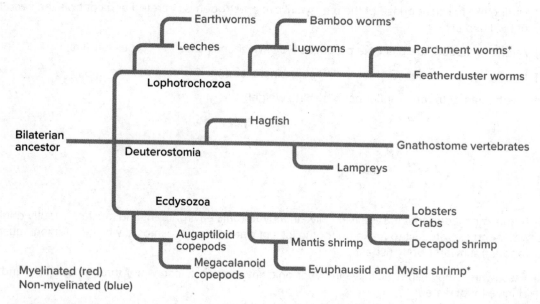

   Data obtained from: Hartline, D. K., and D. R. Colman. 2007. Rapid conduction and the evolution of giant axons and myelinated fibers. Current Biology 17(1): R29-R35.

   (a) **Describe** the function of a myelinated axon. How is this different from a non-myelinated axon?

   (b) **Provide a hypothesis** as to why some invertebrates may have myelinated axons and others do not. **Identify** any patterns in the phylogenic tree you may see.

43 Behavioral Ecology

FOLLOWING *the* BIG IDEAS

AP Essential Knowledge	Chapter Section
BIG IDEA 2 **2.C.2** Organisms respond to changes in their external environment	43.2, 43.3, 43.4
2.E.1 Timing and coordination of specific events are necessary for the normal development of an organism, and these events are regulated by a variety of mechanisms.	43.2
2.E.2 Timing and coordination of physiological events are regulated by multiple mechanisms.	43.2, 43.4
2.E.3 Timing and coordination of behavior are regulated by various mechanisms and are important in natural selection.	43.1, 43.3, 43.4
BIG IDEA 3 **3.D.1** Cell communication processes share common features that reflect a shared evolutionary history.	43.3
3.E.1 Individuals can act on information and communicate it to others.	43.1, 43.2, 43.3, 43.4
3.E.2 Animals have nervous systems that detect external and internal signals, transmit and integrate information, and produce responses.	43.1

CHAPTER OVERVIEW

Behavior allows organisms to adjust to individual and environmental change. Behavior can either be innate or learned. Behavior includes communication. Communication is critical for organisms to function within a community, and there are many forms of communication which have evolved. Behavioral ecology is the study of how natural selection shapes behavior, including communication.

43.1 Inheritance Influences Behavior

Essential Knowledge covered
2.E.3: Timing and coordination of behavior are regulated by various mechanisms and are important in natural selection.
3.E.1: Individuals can act on information and communicate it to others.
3.E.2: Animals have nervous systems that detect external and internal signals, transmit and integrate information and produce responses.

Recall It

Behavior is any action that can be observed or described within an organism. Most behaviors have a genetic basis, and the results of several studies support this hypothesis. These studies include, nest building in lovebirds, food choices in garter snakes, as well as twin studies in humans. One study found that there was actually a gene, gene *fosB*, that controls maternal nurturing behavior in mice.

Review It

Define *behavior*.

43.1 Inheritance Influences Behavior *continued*

Explain the phrase "nurture versus nature."

Use It

How do marine snails demonstrate that behavior has genetic basis?

The graph below demonstrates the number of tongue flicks performed by two populations of garter snakes when exposed to slugs. Snakes use tongue flicks to "smell" their prey. Hypothesize a reason why coastal snakes may eat slugs and inland snakes do not.

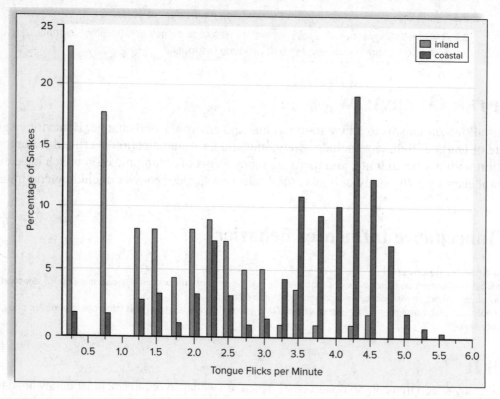

If the behavior of eating slugs in these garter snakes are indeed the result of a genetic basis, what would you expect the offspring of a coastal and inland snake to eat?

43.2 The Environmental Influences Behavior

Essential Knowledge covered
2.C.2: Organisms respond to changes in their external environment.
2.E.1: Timing and coordination of specific events are necessary for the normal development of an organism, and these events are regulated by a variety of mechanisms.
2.E.2: Timing and coordination of physiological events are regulated by multiple mechanisms.
3.E.1: Individuals can act on information and communicate it to others.

Recall It

While most behavior has a genetic basis, some behaviors are influenced by the environment. Fixed action patterns are responses elicited by a sign stimulus but these can be modified by learning. Learning is a durable change in behavior brought about by experience. There are many types of learning. Solving a problem without having prior learning experience about the situation is called insight learning. Imprinting is a form of learning in which a young animal develops an association with the first moving object it sees. The change in behavior that involves an association between two events is known as associated learning. Two forms of associated learning are classical condition and operant conditioning.

Review It

Use the words *orientation*, *navigation*, and *migration* in a cohesive sentence.

Provide an example of the following types of learning.

Learning type	Example
habituation	
operant conditioning	
insight learning	
classical conditioning	
imprinting	
associative learning	

Use It

How do social interactions assist white-crowned sparrows learning how to sing?

43.2 The Environmental Influences Behavior *continued*

Create a flow chart to describe, in general terms, Pavlov's classical conditioning experiment.

43.3 Animal Communication

Essential Knowledge covered
2.C.2: *Organisms respond to changes in their external environment.*
2.E.3: *Timing and coordination of physiological events are regulated by multiple mechanisms.*
3.D.1: *Cell communication processes share common features that reflect a shared evolutionary history.*
3.E.1: *Individuals can act on information and communicate it to others.*

Recall It

Communication is defined by an action from a sender that has the ability to influence the behavior of the receiver. There are many types of communication: (1) chemical, (2) auditory, (3) visual, and (4) tactile. Each type of communication has advantages and disadvantages, but all forms of communication are important for animals to be able to be social and pass messages to one another.

Review It

Determine the following types of communication.

Communication	Example
	a baboon opens its mouth really wide to warn another baboon to leave
	an ant marks its trail with a pheromone
	a bee performs a waggle dance to lead itself to food
	a bird calls an alarm to warn another bird

Define *society*.

List four reasons why organisms communicate.

Use It

Describe two advantages of chemical communication.

43.3 Animal Communication *continued*

Provide an example of visual communication.

43.4 Behaviors that Increase Fitness

Essential Knowledge covered
2.C.2: Organisms respond to changes in their external environment.
2.E.2: Timing and coordination of physiological events are regulated by multiple mechanisms.
2.E.3: Timing and coordination of behavior are regulated by various mechanisms and are important in natural selection.
3.E.1: Individuals can act on information and communicate it to others.

Recall It

Behavioral ecology is the study of how natural selection shapes behavior. The behaviors discussed in this section include: (1) territoriality, (2) reproductive strategies, (3) social behaviors, and (4) altruistic behaviors. Traits or behaviors that foster reproductive success and overall fitness are usually advantageous, although disadvantages do arise.

Review It

Fill out the missing the term or definition on the chart below.

Term	Definition
altruism	
	an animal helps another animal but gains no immediate benefit but is repaid at some later time
inclusive fitness	
	the adaptation to the environment due to reproductive success of the individual's relatives
polyandrous	
	the study of how natural selection shapes behavior
	a pair bonds to produce an offspring and both care for the young
polygamous	

Use It

Why would a shore crab prefer a medium sized mussel to eat instead of a large one?

43.4 Behaviors that Increase Fitness *continued*

The male bowerbird has developed a flamboyant display in order to make the female notice him. If the male becomes too intense in his display, however, what happens?

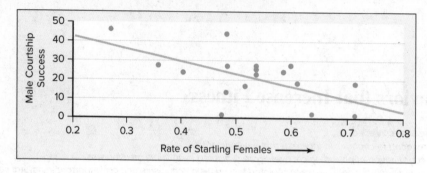

AP **CHAPTER SUMMARY**

Summarize It

How do genetics and the environment work together to influence both innate and learned behaviors?

What are examples of strategies that animals use to communicate information, and how do these strategies increase survival and reproductive success, i.e., fitness?

Describe two ways in which altruism is beneficial.

FOLLOWING *the* BIG IDEAS

	AP Essential Knowledge	Chapter Section
BIG IDEA 2	**2.D.1** All biological systems from cells to organisms to populations, communities and ecosystems are affected by complex biotic and abiotic interactions involving exchange of matter and free energy.	44.1, 44.2, 44.3, 44.4, 44.5
	2.D.3 Biological systems are affected by disruptions to their dynamic homeostasis.	44.4
BIG IDEA 4	**4.A.5** Communities are composed of populations or organisms that interact in complex ways.	44.2, 44.3, 44.4, 44.5, 44.6
	4.A.6 Interactions among living systems and with their environment result in the movement of matter and energy.	44.6
	4.B.3 Interactions between and within populations influence patterns of species distribution and abundance.	44.3, 44.4

CHAPTER OVERVIEW

Ecology is the study among all organisms and their interactions with their physical environment. A population is defined as all the organisms belonging to the same species within an area at the same time. Populations are impacted by both biotic and abiotic factors. Population ecology is the field which models the impact of these factors on population growth. There are many factors which ultimately limit the growth of population.

44.1 Scope of Ecology

Essential Knowledge covered
2.D.1: All biological systems from cells and organisms to populations, communities and ecosystems are affected by complex biotic and abiotic interactions involving exchange of matter and free energy.

Recall It

Ecology is an enormous field of biology which focuses on the interactions of organisms and their environments. The place where an organism lives is known as its habitat. The same species within a habitat is known as a population. All of the populations at a particular locale is called a community. An ecosystem is the communities along with abiotic factors. Encompassing of all the soil, water, and air is called the biosphere. It is the ultimate goal of ecologists to explain and predict the distribution and abundance of organisms in the biosphere.

44.1 Scope of Ecology *continued*

Review It

Place the following ecological levels in order from least complex to most complex: *community, biosphere, population, habitat, ecosystem*. Provide a definition and example for each.

Level	Definition	Example

Use It

Describe how ecology and evolution are intertwined.

44.2 Demographics of Populations

Essential Knowledge covered

2.D.1: All biological systems from cells and organisms to populations, communities and ecosystems are affected by complex biotic and abiotic interactions involving exchange of matter and free energy.

4.A.5: Communities are composed of populations of organisms that interact in complex ways.

Recall It

The statistical study of a population is called demography. Studies can be performed on the number of individuals per unit area or on population density, the pattern of dispersal of individuals. Population distribution describes how the population is growing through the rate of natural increase. Factors that are used to determine population distribution include resources and limiting factors, such as environmental aspects that determine where organisms live. The highest possible rate of natural increase for a population is known as its biotic potential. The probability of how long an individual will live to a certain age is known as survivorship. The numbers of individuals alive in each generation of a population can be mapped in an age structure diagram.

Review It

List four important resources that humans need to survive.

Identify two limiting factors which may reduce a population's potential reproduction.

Use It

Calculate the rate of natural increase for a population of 2000 that has 50 births per year and 15 deaths. Recall that the rate of natural increase (*r*) is determined by the number of births per year minus the number of deaths per year divided by the number of individuals in a population.

Identify if the statement pertains to survivorship curve I, II, or III shown on the graph below.

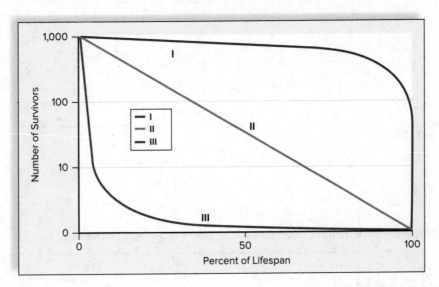

Most large mammals follow this type of survivorship curve.	
Death is unrelated to age.	
Death does not come until near the end of the lifespan.	
Most individuals die very young.	

44.3 **Population Growth Models**

Essential Knowledge covered
2.D.1: *All biological systems from cells and organisms to populations, communities and ecosystems are affected by complex biotic and abiotic interactions involving exchange of matter and free energy.*
4.A.5: *Communities are composed of populations of organisms that interact in complex ways*
4.B.3: *Interactions between and within populations influence patterns of species distribution and abundance.*

Recall It

There are two working models for population growth. One is based on semelparity, where the other is based off the pattern of iteroparity. Semelparous organisms can best be modeled by an exponential growth curve. This curve has two phases: a lag phase and an exponential growth phase. When limiting environmental factors come into play, logistic growth curves are a better fit, and include a deceleration phase and stable equilibrium phase, in addition to the lag and exponential growth phases. The maximum number of individuals of a given species a community can support is known as its carrying capacity (K).

Review It

Define the models, curves, and factors of population growth.

Term	Definition
exponential growth	
semelparity	
logistic growth	
iteroparity	
carrying capacity (*K*)	

Use It

The graph below shows the number of yeast cells per hour growing in a beaker.

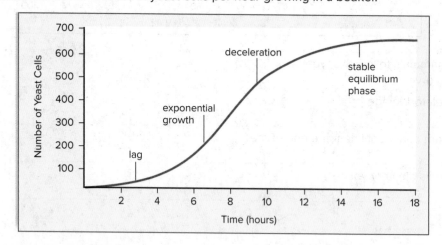

What is this type of growth called?

Describe what happens to yeast population during the four phases of its growth.

What types of factors are necessary for this type to growth to continue?

Plot the following data on the line graph.

Generation	Population
1	4
2	4
3	10
4	20
5	40

What is this type of growth curve called?

What are the two phases of this type of growth curve?

What types of factors are necessary for this type to growth to continue?

44.4 Regulation of Population Size

Essential Knowledge covered
2.D.1: All biological systems from cells and organisms to populations, communities and ecosystems are affected by complex biotic and abiotic interactions involving exchange of matter and free energy.
2.D.3: Biological systems are affected by disruptions to their dynamic homeostasis.
4.A.5: Communities are composed of populations of organisms that interact in complex ways.
4.B.3: Interactions between and within populations influence patterns of species distribution and abundance.

Recall It

Abiotic and biotic conditions play a role in regulating population size. Abiotic factors, such as droughts, floods, and forest fires, are density-independent factors, meaning the intensity of the effect does not increase with increased population density. Biotic factors, such as disease or predation, are density-dependent factors; the percentage of the population affect does increase as the population increases.

44.4 Regulation of Population Size *continued*

Review It

Provide an example of the following terms.

Term	Example
density-independent factors	
density-dependent factors	
competition	
predation	

Identify if the statement pertains to a density-dependent (D) or density-independent (I) factor.

The intensity of the effect does not increase with the increased population density.	
It is usually a biotic factor	
It is usually an abiotic factor	
The intensity of the effect does increase with the increased population density	

Use It

A volcano erupts and destroys an entire village nearby. Is the volcano eruption a density-independent or dependent factor? Explain your answer.

44.5 Life History Patterns

Essential Knowledge covered

2.D.1: *All biological systems from cells and organisms to populations, communities and ecosystems are affected by complex biotic and abiotic interactions involving exchange of matter and free energy.*

4.A.5: *Communities are composed of populations of organisms that interact in complex ways*

Recall It

The number of births per reproductive event, the age of reproduction, the lifespan, and the probability of living the entire lifespan are all factors taken into account when determining how populations change. It is thought that organisms follow a life history pattern that is more like an *r*-strategist or *K*-strategist. Opportunistic species are more likely to be like an *r*-strategist; small in size with a short lifespan, fast to mature, with many offspring. *K*-strategist are usually larger, with long lifespans, slow maturation rates, and have few offspring which they provide much care.

44.5 Life History Patterns *continued*

Review It

Describe the populations with *K*- or *r*-selection.

K-selection	*r*-selection

Identify if the statement pertains to populations with *K*- or *r*-selection.

Often good at colonizing new habitats	
Has a fairly long lifespan.	
Many offspring die before reproducing.	
Favor stable, predictable environments.	

Use It

Which life history pattern do humans most closely follow? Explain your answer.

Are all organisms either *K*- or *r*-strategists?

44.6 Human Population Growth

Essential Knowledge covered
4.A.5: Communities are composed of populations of organisms that interact in complex ways.
4.A.6: Interactions among living systems and with their environment result in the movement of matter and energy.

Recall It

The number of humans on Earth is over 7.1 billion people and growing. There is much concern about the amount of natural resources needed to support this population size. Countries are divided into more-developed and less-developed countries which have distinct patterns of growth. Age distributions show that more-developed countries have leveled off in growth; whereas, less-developed countries are still rapidly increasing.

Review It

Describe three problems a less-developed country might face if populations continue to climb in these regions.

44.6 Human Population Growth *continued*

Use It

People living in more-developed countries face a second type of overpopulation. Use the environmental impact equation to describe this second type of over-population.

Summarize It

Describe why a fishing company would want to monitor the carrying capacity of its fish populations.

Describe how competition for food could control population growth.

How do environmental factors, including energy availability, affect the density and distribution pattern of a population?

45 Community and Ecosystem Ecology

FOLLOWING *the* BIG IDEAS

AP Essential Knowledge	Chapter Section
BIG IDEA 1 **1.A.3** Evolutionary change is also driven by random processes.	45.2
BIG IDEA 2 **2.A.1** All living systems require constant input of free energy.	45.3
2.A.3 Organisms must exchange matter with the environment to grow, reproduce and maintain organization.	45.3
2.D.1 All biological systems from cells to organisms to populations, communities and ecosystems are affected by complex biotic and abiotic interactions involving exchange of matter and free energy.	45.1, 45.3
2.E.3 Timing and coordination of behavior are regulated by various mechanisms and are important in natural selection.	45.1
BIG IDEA 4 **4.A.5** Communities are composed of populations or organisms that interact in complex ways.	45.1, 45.3
4.A.6 Interactions among living systems and with their environment result in the movement of matter and energy.	45.1, 45.3
4.B.3 Interactions between and within populations influence patterns of species distribution and abundance.	45.1, 45.4
4.B.4 Distribution of local and global ecosystems changes over time.	45.2, 45.4

CHAPTER OVERVIEW

In an ecosystem, biotic and abiotic components interact. There are various types of interactions among the populations of a community, as there are interactions with the nonliving component of an ecosystem. Following the flow of energy in an ecosystem is critical in understanding its structure. This includes knowing who eats who or who eats what. The biogeochemical cycles of critical chemicals necessary for life are also fundamental in understanding ecosystem ecology.

45.1 Ecology of Communities

Essential Knowledge covered
2.D.1: All biological systems from cells and organisms to populations, communities and ecosystems are affected by complex biotic and abiotic interactions involving exchange of matter and free energy.
2.E.3: Timing and coordination of behavior are regulated by various mechanisms and are important in natural selection.
4.A.5: Communities are composed of populations of organisms that interact in complex ways.
4.A.6: Interactions among living systems and with their environment result in the movement of matter and energy.
4.B.3: Interactions between and within populations influence patterns of species distribution and abundance.

Recall It

As discussed in Chapter 44, populations are parts of communities. Communities can be described through species richness and through species diversity. There are many different types of interactions which occur in populations, and these interactions help shape a community and drive evolution. Competition is one such interaction. The competitive exclusion principle states that no two species can indefinitely occupy the same niche at the same time. Predator-prey relationships and mutualism are other forms of interactions which drive community structure.

45.1 Ecology of Communities *continued*

Review It

Which is more diverse: a pond which has 30 wood frogs, 20 green frogs, and 10 bull frogs, or a pond that has 80 wood frogs, 2 green frogs, and 5 bullfrogs. Justify your answer.

Explain the difference between Batesian and Müllerian mimicry.

Use It

Compare and contrast camouflage and mimicry.

In a famous niche partitioning study, Joseph Cornell studied two species of barnacles living on the same rock. Using the illustration provided below, describe the niches and characteristics that allow two species to occupy the same rock.

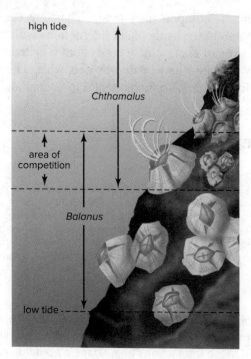

45.2 Community Development

Essential Knowledge covered
1.A.3: Evolutionary change is also drive by random processes.
4.B.4: Distribution of local and global ecosystems change over time.

Recall It

Communities grow and change over time. Ecological succession is the change within a community where species replace other species as time goes on. Primary succession is the initital formation of a community; whereas, secondary succession occurs after a disturbance occurs in an ecosystem and there is a progressive change in species. The first producers to colonize a community are called pioneer species. It is thought that succession in a particular area will eventually lead to a stable and mature community, called a climax community.

Review It

Identify the pioneer species and climax community in the drawing of ecological succession above.

Distinguish between primary and secondary succession.

Use It

Describe an event which happens on a short timescale and an event which happens over a long timescale which leads to ecological change.

Compare and contrast the three models of succession.

45.3 **Dynamics of an Ecosystem**

Essential Knowledge covered
2.A.1: *All living systems require constant input of free energy.*
2.A.3: *Organisms must exchange matter with the environment to grow, reproduce, and maintain organization.*
2.D.1: *All biological systems from cells and organisms to populations, communities and ecosystems are affected by*
4.A.5: *Communities are composed of populations of organisms that interact in complex ways.*
4.A.6: *Interactions among living systems and with their environment result in the movement of matter and energy.*

Recall It

In an ecosystem, the biotic organisms are categorized by how they obtain their nutrients; as either autotrophic, heterotrophic, or detritivores. Energy flows between these biotic organisms in the form of a food web. As discussed very early on in this textbook, the initial form of energy comes from the sun. As energy flows between each organism in a food web, energy is lost. The arrangement of species in a food wed is known as a food chain and are organized into trophic levels. While energy is lost in these food webs, the nutrients from abiotic sources are not; they are circulated through biogeochemical cycles. The water, carbon, phosphorus, nitrogen, and sulfur cycle are major biogeochemical cycles organisms rely on, and each can be impacted by human activities.

Review It

List the four main biogeochemical cycles.

Using the following list of organisms, draw a grazing food chain and an ecological pyramid.

Grass, hawk, rabbit

45.3 Dynamics of an Ecosystem *continued*

Use It

Imagine you are on a grassy plain in Africa. Provide some ideas of the types of producers, consumers, and decomposers which might be in that ecosystem. Where does the energy they need to reproduce and survive come from?

	Producers	Consumers	Decomposers
Examples			
Energy Source			

Describe two ways that humans can interfere with the water cycle, and how this may affect other organisms.

Compare and contrast the phosphorus and nitrogen cycle.

45.4 Ecological Consequences of Climate Change

Essential Knowledge covered
4.B.3: Interactions between and within populations influence patterns of species distribution and abundance.
4.B.4: Distribution of local and global ecosystems changes over time.

Recall It

Earth's climate has changed dramatically over the course of our planet's history. These changes were largely caused by small variations in Earth's orbit around the sun. The changes currently occurring are unlike those previous: they are happening rapidly, and they are happening because of human activity. These activities include production of greenhouse gases and deforestation. The severity of the consequences of climate change depend on whether or not we make changes to our lifestyles to curb our impact on the climate.

45.4 Ecological Consequences of Climate Change *continued*

Review It
List two greenhouse gases and their primary sources.

Use It
Explain the difference between the greenhouse gas effect, climate change, and global warming.

Identify two potential consequences of climate change. How will these consequences intensify if nothing is done to curb climate change?

AP CHAPTER SUMMARY

Summarize It
How do interactions among species and the ecosystem organize a community?

Bats and bluebirds both live near fields and both eat the same types of moths and beetles. How are they able to occupy the same habitat?

Describe what happens to a pond when runoff from fertilizer applied to a lawn introduces too much nitrogen into the system. What biogeochemical cycles are affected, and how does this impact the flow of energy in this ecosystem?

FOLLOWING *the* BIG IDEAS

AP Essential Knowledge	Chapter Section
BIG IDEA 4 **4.A.6** Interactions among living systems and with their environment result in the movement of matter and energy.	46.1, 46.2, 46.3
4.B.4 Distribution of local and global ecosystems changes over time.	46.3

CHAPTER OVERVIEW

Climate is the prevailing weather conditions in a particular region and determines the producer at the base of a biome. Solar radiation and topography play key roles in determining the climate of a region. These regions are then categorized as particular biomes. Each biome has a particular type of plants and animals that are well adapted to living in the environmental conditions. Also mentioned in this chapter, are the aquatic environments. Aquatic environments are categorized as freshwater or saltwater systems.

46.1 Climate and the Biosphere

Essential Knowledge covered
4.A.6: Interactions among living systems and with their environment result in the movement of matter and energy.

Recall It

Climate is the prevailing weather conditions in a particular region and is affected by rainfall and temperature. Solar radiation and topography influence rainfall and temperature. The spherical nature of the Earth leads to variation in solar radiation, as does its tilt and rotation. Topography is the physical feature of the landscape. Mountains, in particular, affects the climate through forming rain shadows.

Review It

Fill in the term or its definition associated with climate.

Term	Definition
rain shadow	
	the prevailing weather conditions in particular regions
	the physical features of the land
monsoon	

Describe two factors which dictate the climate of a region.

46.1 Climate and the Biosphere *continued*

Use It

If it is close to the autumnal equinox, and arctic winds have been blowing east, what might the climate be like in New York City? What might the climate might be like near the Equator?

46.2 Terrestrial Ecosystems

Essential Knowledge covered
4.A.6: Interactions among living systems and with their environment result in the movement of matter and energy.

Recall It

A biome has a specific range of environmental conditions within a particular geological location. These biomes, in turn, support specific plants and animals that have adapted to survive in this particular climate. There are many types of biomes on Earth. For this section, use your textbook to identify the major characteristics of each biome listed and the vegetation found there. Do you see any trends of interest?

Review It

Characterize the climate and vegetation of the following biomes.

Biome	Climate	Vegetation Type
alpine tundra		
arctic tundra		
chaparral		
deserts		
grasslands		
savannas		
shrublands		
taiga		
temperate deciduous forest		
temperate grasslands		
temperate rain forest		
tropical rain forest		

46.2 Terrestrial Ecosystems *continued*

Use It

Explain, in terms of energy and resources, the variation in vegetation types found between the different biomes.

46.3 Aquatic Ecosystems

Essential Knowledge covered
4.A.6: Interactions among living systems and with their environment result in the movement of matter and energy.
4.B.4: Distribution of local and global ecosystems change over time.

Recall It

Aquatic ecosystems are classified as either saltwater or freshwater. They are further identified as wetlands, which has its own system of classifications, lakes, costal ecosystems, and oceans. Aquatic ecosystems support the majority of primary productivity on Earth. Aquatic ecosystems are often highly diverse, and many terrestrial ecosystems depend on the resources and habitats aquatic ecosystems can provide. Meteorological, geological, or anthropomorphic events can impact all ecosystems.

Review It

Identify the following terms based on their description.

Description	Term
Areas of biological abundance found in warm, shallow tropical waters	
Where fresh water and salt water mix	
The concentration of pollutants as they move up the food chain	
The region of the shoreline that lies between the high and low tidal marks	
Offshore winds cause cold nutrient-rich waters to rise and displace warm nutrient-poor water	
A place where seawater is heated to about 350°C and percolates through the cracks at the bottom of the ocean	
The lack of upwelling which causes stagnation and climate patterns to change	

List the four zones of a lake, and provide an example of what lives in each zone.

46.3 Aquatic Ecosystems *continued*

Use It

List three ways that wetlands benefit humans.

Describe how ocean currents can change the climate and nutrients of an aquatic habitat.

AP CHAPTER SUMMARY

Summarize It

What environmental factors determine the location and nature of terrestrial and aquatic biomes and ecosystems?

FOLLOWING *the* BIG IDEAS

AP Essential Knowledge	Chapter Section
BIG IDEA 1 **1.A.2** Natural selection acts on phenotypic variations in populations.	47.1, 47.2
1.C.1 Speciation and extinction have occurred throughout the Earth's history.	47.3
1.C.3 Populations of organisms continue to evolve.	47.2
BIG IDEA 4 **4.A.5** Communities are composed of populations or organisms that interact in complex ways.	47.2, 47.3
4.A.6 Interactions among living systems and with their environment result in the movement of matter and energy.	47.3
4.B.3 Interactions between and within populations influence patterns of species distribution and abundance.	47.3
4.B.4 Distribution of local and global ecosystems changes over time.	47.3, 47.4
4.C.3 The level of variation in a population affects population dynamics.	47.1, 47.2
4.C.4 The diversity of species within an ecosystem may influence the stability of the ecosystem.	47.1, 47.2, 47.3

CHAPTER OVERVIEW

Human activities are causing major negative impacts on ecosystems. As a result, many species are going extinct. Conservation biology has emerged as a field of biology that aims to study biodiversity and how we can best conserve natural resources. There is great value in maintaining biodiversity, and conversation techniques are being developed and tested in order how to figure out how best to do this.

47.1 Conservation Biology and Biodiversity

Essential Knowledge covered
1.A.2: Natural selection acts on phenotypic variations in populations.
4.C.3: The level of variation in a population affects population dynamics.
4.C.4: The diversity of species within an ecosystem may influence the stability of the ecosystem.

Recall It

The goal of conservation biology is to study the impact of biodiversity with the goal of conserving natural resources. Biodiversity if the variety of life on Earth. Biodiversity can be thought of in terms of genetic diversity, community diversity, and landscape diversity. Some regions of the world are biodiversity hotspots; meaning they contain a large concentration of different species. Identifying these locations, as well as diversity at each level mentioned above, allows scientists to identify endangered and threatened species, and threats to biodiversity.

Review It

List the three levels of biodiversity.

Why is a small, isolated population more likely to become extinct than a larger, more connected population?

Use It

Why do conservation biologists study genetic, community, and landscape diversity as well as just counting the numbers of a particular species?

How can it be detrimental to only try to conserve a charismatic species? Give an example of such an approach.

47.2 Value of Biodiversity

Essential Knowledge covered
1.A.2: Natural selection acts on phenotypic variations in populations.
1.C.3: Populations of organisms continue to evolve.
4.A.5: Communities are composed of populations of organisms that interact in complex ways.
4.C.3: The level of variation in a population affects population dynamics.
4.C.4: The diversity of species within an ecosystem may influence the stability of the ecosystem.

Recall It

Biodiversity has both direct and indirect value. Direct value is a concrete value, such as medicinal value, agricultural value, and consumptive use value. Indirect value is a little more complex as ecosystems also provide services that do not have a measureable economic value. This includes biogeochemical cycles, waste recycling, prevention of soil erosion and flooding, regulation of climate, and ecotourism. Biodiversity appears to be fundamental to stability in ecosystems.

47.2 Value of Biodiversity *continued*

Review It

Explain the difference between the direct and indirect value of biodiversity.

Differentiate between direct value and indirect value. Provide two examples of each.

Provide an example of how biodiversity is critical in each of the following ecological processes:

Process	Benefit from Biodiversity
biogeochemical cycles	
waste recycling	
soil stability	
climate	

Use It

Why is it important to maintain genetic diversity with a crop species?

How might a high degree of biodiversity help an ecosystem function more efficiently?

47.3 Causes of Extinction

Essential Knowledge covered
1.C.1: *Speciation and extinction have occurred throughout the Earth's history.*
4.A.5: *Communities are composed of populations of organisms that interact in complex ways.*
4.A.6: *Interactions among living systems and with their environment result in the movement of matter and energy.*
4.B.3: *Interactions between and within populations influence patterns of species distribution and abundance.*
4.C.4: *The diversity of species within an ecosystem may influence the stability of the ecosystem.*

47.3 Causes of Extinction *continued*

Recall It

Human are driving many species to extinction. Habitat loss is occurring all ecosystems, much of this is due to deforestation and construction of roads and cities. Humans have also introduced and spread "exotic" or new species into different habitats, where they lack natural predators. Human pollution has come in many forms from acid deposition, eutrophication, and organic chemicals and affected the health and lives of living organisms. Human activities have led to an acceleration in climate change, which threatens species around the globe. Finally, humans directly remove species in abundance through exploitation for food or game.

Review It

Define the following threats to biodiversity.

Threat	Definition
habitat loss	
overexploitation	
pollution	
exotic species	
climate change	

List two species humans have accidently transported from one environment to another which threaten biodiversity.

Use It

How can introducing a new species to an environment it is not native to affect the other species living there?

Describe three types of pollutants and how they negatively affect biodiversity.

47.3 Causes of Extinction *continued*

In the marine ecosystem o of the coast of California, sea otters are recovering from a genetic bottleneck caused by overharvesting by the fur trade. Sea otters eat sea urchins. Urchins are a main grazer on kelp, which provides a habitat for commercial and recreational fisheries. Seals and sea lions feed on fish. Orcas will eat otters but prefer sea lions.

Draw a diagram of the coastal food web. What will happen to the coastal fisheries the sea otter population is reduced by disease, as is a risk for populations in a genetic bottleneck?

47.4 Conservation Techniques

Essential Knowledge covered
4.B.4: Distribution of local and global ecosystems change over time.

Recall It

A species that plays a fundamental role in the operation of a community is known as a keystone species. Keystone species are different from flagship species, which are species that evoke a strong emotion response in humans. In order to help conserve species, it is important to understand that some species now live in metapopulations, as a result of habitat fragmentation Within these populations, there is a source population, in which one population produces an abundance of individuals, which may migrate to a sink population. Fragmentation leads to an increase in the edge effect. Habitat preservation and restoration as are equally as important in protecting individual species in conservation biology.

Review It

Why are keystone species important to an ecosystem? Provide one example of a keystone species.

Describe the three key principals of restoration ecology.

47.4 Conservation Techniques *continued*

Use It

Refer to Figure 47.13 on page 909 of your textbook. Describe what happens to viable habitat when an environment becomes fragmented into smaller and smaller pieces.

AP CHAPTER SUMMARY

Summarize It

How are human activities contributing to the endangerment and possible extinction of other species? What responsibility do we have for maintaining the Earth's biodiversity?

What is the value of biodiversity to humans? Why does it matter if species become extinct?

UNIT EIGHT: BEHAVIOR AND ECOLOGY

Chapter 43 Behavioral Ecology • Chapter 44 Population Ecology • Chapter 45 Community and Ecosystem Ecology • Chapter 46 Major Ecosystems of the Biosphere • Chapter 47 Conservation of Biodiversity

Multiple Choice Questions

Directions: For each question or incomplete statement choose one of the four suggested answers or completions listed below.

1. The following chart shows the duration of hibernation of woodchucks from three different states:

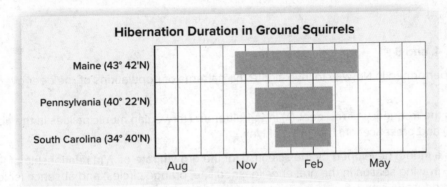

Data obtained from: Williams, C. T., et al. 2014. Phenology of hibernation and reproduction in ground squirrels: integration of environmental cues with endogenous programming. Journal of Zoology 292 (2): 112–124.

Which conclusive statement can be made by looking at this graph?

(A) Differences in snow cover are associated with the timing of spring emergence in woodchucks.

(B) Differences in hibernation time in woodchucks is influenced by genetics.

(C) Differences in hibernation time in woodchucks is influenced by latitude.

(D) Differences in soil temperatures are associated with the timing of spring emergence in woodchucks.

2. Honeybees have a unique form of communication called the "waggle dance," which indicates the distance and direction of a food sources to other bees. If a commonly used pesticide is shown to disrupt waggle dance behavior in honeybees, what is a likely short-term outcome?

(A) Honeybees will become lost looking for food.

(B) The colony will lose nutrients and fitness.

(C) Honeybees will find another way to look for food sources.

(D) The colony will become resistant to the pesticide.

Questions 3, 4, and 5

A study was performed in Norway to investigate the patterns of populations of roe deer over a nine-year period.

Part a. shows the frequency of roe deer (λ) in 144 different Norwegian municipalities in the absence (solid orange) and presence (dashed gray) of lynx.

Part b. shows a model developed by the scientists of the growth rate of λ in relationship of the varying length of the growing season in the presence (dashed line, orange circles) and absence (sold line, grey circles) of reproductive lynx.

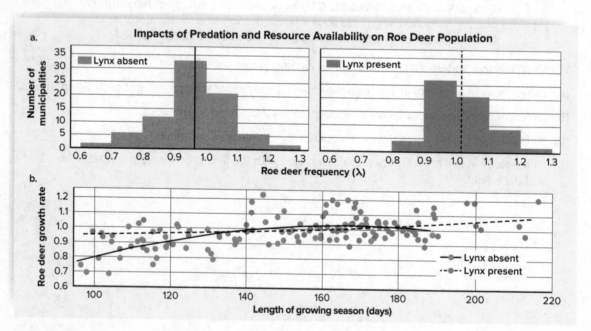

Data obtained from: Melis, Claudia, et al. 2010. Roe deer population growth and lynx predation along a gradient of environmental productivity and climate in Norway.

3. Based on these data, the relationship between the roe deer and the lynx can best be described as:

(A) Symbiont

(B) Predator-Prey

(C) Parasitic

(D) Commensal

4. There was a wide variation in the climatic conditions in the 144 municipalities studied. Which of the following is true about regions with shorter growing seasons?

(A) Regions with shorter growing seasons have less primary productivity.

(B) Regions with shorter growing seasons have milder climates.

(C) Regions with shorter growing seasons have less lynx.

(D) Regions with shorter growing seasons have more wetlands.

5. According to the scientists' model:

(A) Lynx presence appears to be having a positive impact on roe deer populations.

(B) The length of the growing season has a negative impact on the number of roe deer.

(C) Lynx reproduction increases with increasing day length.

(D) Deer population growth rates will be lower in the municipalities with lynx and shorter growing seasons than in municipalities with mild climatic conditions and/or without lynx.

6. In an ecosystem, energy transformations repeatedly occur and

(A) Energy is always conserved.

(B) Energy derived from the sun eventually dissipates as heat.

(C) All energy is consumed.

(D) Energy is recycled.

Free Response Questions

Directions: Read the questions carefully and completely. Then, plan your answer and write your response in the space provide. Write your answer out in paragraph form.

1. Using a diagram, **explain** how the Edge Effect reduces the amount of habitat of a forest ecosystem, and **hypothesize** how the Edge Effect might affect a population of song birds.

2. The data from a study performed in four national parks in the U.S. Rocky Mountains found that parks that had more wetlands created or modified by beavers had higher colonization rates of different types of frogs and toads.

   (a) **Describe** the influence of the beaver on the U.S. Rocky Mountain national parks using ecological terms.

   (b) If you were a park ranger and wanted to aid in protecting endangered amphibians in the National parks, what management plan would you suggest implementing?